The Empire Express

The story of the U.S. Navy PV squadrons' aerial strikes against the Japanese Kuriles during WW II.

CHARLES L. SCRIVNER

PUBLISHED BY

P.O. BOX 33
TEMPLE CITY, CALIF. 91780

INTRODUCTION AND ACKNOWLEDGMENT

The chain of Aleutian Islands belonged to the United States ever since the purchase of the Alaska Territory from Russia in 1867. As Secretary of State, William H. Seward (R.) pressed Congress to purchase the land as an investment for an unheard of sum of $7,200,000. After much wrangling it was so done but the bleak, barren land, far removed from the States themselves, was quickly dubbed "Seward's folly." For decades the land lay unmolested, a useless formation of volcanic rock, ash, bitter cold weather and frozen tundra. The area was virtually undeveloped until the 1920s and of little consequence militarily except as a remote communications outpost. In the early 1930s, airpower advocate Billy Mitchell and military historian Homer Lea foresaw the importance of Alaska and the Aleutian stepping stones as of strategic value to America — as well as a potential enemy. Their alarm went unheeded.

In December 1941, when the Japanese struck at Pearl Harbor, Hawaii, the Aleutians were uninhabited except for a few Aleuts and a small Army and Navy contingent at Dutch Harbor. Before that fateful Sunday was over, the United States and Japan were at war. The U.S. Navy was stunned and for months thereafter was incapable of effectively challenging the enemy. Admiral Yamamoto's Imperial Fleet did roam the Pacific at will. In their westward island hopping across the Pacific, Yamamoto's schedule called for the conquest of a small but highly valuable island in mid-Pacific called Midway. The date was set for the first week in June. At the same time a prophetic diversionary tactic was devised.

Admiral Yamamoto sent Admiral Hosogaya, with two small aircraft carriers, the **Junyo** and **Ryujo** and supporting ships, north, as their "Aleutian Screening Force". On 3-4 June '42 the task force bombed Dutch Harbor. Landing of troops was not planned but it kept American forces occupied and in turmoil over the main intent of Yamamoto. Fortunately U.S. Naval intelligence was not fooled and they were ready for the Midway battle.

However, the Japanese Northern Area Force was stronger and more definitive than expected. On 7 June they were able to occupy Attu unopposed and Kiska a week later. For eleven months the Japanese tried to build up airfields and defenses on the islands. The Army and Navy tried equally hard to dislodge them with aerial bombing and ship shellings. In the end neither made progress — the weather was the only victor.

Following the devastating defeat of the Imperial Navy at Midway, both adversaries made serious reassessments of the Aleutians, but it wasn't until mid-1943 that the American forces were strong enough to attempt the reoccupation of Attu and Kiska. On 11 May '43, the U.S. Army's 7th Division stormed ashore on Attu. It cost the lives of 700 Americans and it wounded 1,500. Most of the 2,500 enemy troops were killed in a mad suicidal banzai charge while small groups who survived committed hara-kiri. Expecting an even longer and bloodier battle to retake Kiska, 35,000 American and Canadian troops invaded the jagged island on 17 Aug. '43. They had a surprise, no resistence was met, the enemy had evacuated some 19 days earlier, slipping through the Navy's screen.

With the capture of the American islands, the U.S. Navy immediately set about establishing an air base on Attu, the western most island suitable for operations by land based aircraft. Casco Field was built and Navy bomber squadrons moved in. Their job was to take the offensive — to strike at the northern end of the Japanese Empire. Thus a special breed of Navy fliers was formed — the Empire Express. This is their story . . . their special niche in aerouantical history.

War stories have a propensity to flower with the passing of the years. Every effort has been made to keep this account historically correct. It is the author's hope that publication of this material will open leads to additional sources of information and to the correction of any errors that may exist.

A host of people have made generous contributions to this presentation of the *Empire Express.* First and foremost, my good friend Captain W. E. "Bill" Scarborough, USN (Ret.), a veteran "PV Driver", who furnished encouragement, enthusiasm, and at times much needed prodding. Special thanks are extended to the following for supplying that main ingredient, first hand help and personal interest:

Admiral James S. Russell, USN (Ret.)
Capt. Robert Feiten, USN (Ret.)
Cdr. M. A. "Butch" Mason, USN (Ret.)
L. A. Patteson
Robert R. Larson
Arthur A. Hoffman, AOCS, USN (Ret.)
Edward P. Vance
Rick Brownlee
Frederick E. Cline, AOC, USNR
Richard M. Hill
Dr. D. C. Allard, Naval History Div. (Air Warfare), Dept. of Navy

Charles L. Scrivner
Independence, Mo. 1976

LIBRARY OF CONGRESS
CATALOG CARD No. 76-17672
ISBN No. 911852-78-6

CHAPTERS

ABBREVIATIONS USED IN THIS BOOK

AA — Anti-Aircraft fire.

ACI — Air Combat Intelligence. The ACI officer was the pre- and post-flight briefer. He maintained files of target info and prepared action reports, operational summaries, etc.

AMM — Aviation Machinist's Mate

AOM — Aviation Ordnanceman

ARM — Aviation Radioman

ASD — Airborne Search and Detection — early WWII aircraft radar equipment which was quite sophisticated and hush-hush. Squadron personnel were introduced to it at secret meetings, an armed Marine guard at the door, and the Executive Officer checking the identity of all who entered.

BETTY — Allied code name for Japanese Mitsubishi G4M-1/G4M-3 Navy land-based twin engined attack bomber.

BOGEY — An unidentified aircraft; usually applied to fighters.

BuNo — U.S. Navy Bureau of Aeronautics' Serial Number for an individual aircraft.

CAVU — Ceiling and Visibility Unlimited.

CINCPAC — Commander in Chief Pacific Fleet — Admiral Nimitz and his staff at Pearl Harbor.

DITCH — To make an emergency crash landing on the water in an airplane which would not float. A last resort measure.

CHIEF — Chief Petty Officer — the highest U.S. Navy enlisted rank.

ENS. — Ensign — equivalent rank to Army Second Lieutenant, lowest commissioned rank.

EXEC — Executive Officer — second in command.

FAW — Fleet Air Wing

FIREWALLED — The throttles, mixture, and propeller controls were advanced to the maximum settings. The term originates from single-engine aircraft, where a firewall separated the pilot from the engine. "The throttle was pushed to the firewall" — WIDE OPEN.

HAMP — Allied code name for Japanese Mitsubishi A6M-3 Navy carrier or land based fighter. A later model of the "Zero" of early WWII fame.

HEDRON — Headquarters Squadron, a supporting unit providing administrative and maintenance services to Patrol Squadrons.

KNOTS — Nautical Miles per Hour, a Nautical Mile being 6076 feet.

LCdr. — Lieutenant Commander — equivalent rank to Army Major.

Lt. — Lieutenant — equivalent to Army Captain.

Lt.(jg) — Lieutenant (Junior Grade) — equivalent to Army First Lieutenant.

OSCAR — Allied code name for Japanese Nakajima Ki-43 Army land based fighter, Type 1 "Hayabusa"

PBY — Consolidated Catalina flying boat. The PBY-5A and -6A were amphibians.

P-BOAT — Navy colloquialism for any patrol plane, — especially seaplanes.

PhoM — Photographer's Mate

PPC — Patrol Plane Commander, Navy designation for the aircraft commander.

PV-1 — The Navy designation for the VENTURA. The first letter, P, is for Patrol. The second, V, is the manufacturer's code letter assigned to the Vega Division of Lockheed Aircraft Corporation.

PV-2 — The HARPOON. A redesign of the PV-1, also manufactured by Lockheed Vega, featuring an additional ten feet of wing span, enlarged bomb bay, and enlarged tail surface area. PV-1 production continued until May 1944, while PV-2 production began in March 1944.

SKIPPER — Commanding Officer.

TOJO — Allied code name for Japanese Nakajima Ki-44 Army land based fighter, Type 2, "Shoki".

USNR — U.S. Naval Reserve. Reserve personnel on active duty, as opposed to a member of the Regular Navy (USN).

VB — Bombing Squadron. The first letter, V, denotes heavier-than-air aircraft. During WWII the Navy still had blimp squadrons which were designated Z for lighter-than-air.

VFR — Visual Flight Rules. Weather conditions permitting flight by reference to ground and horizon, as opposed to IFR, flying by reference to instruments only.

VP — Patrol Squadron.

VPB — Patrol Bombing Squadron.

WILLI-WAW — A sudden violent gust of cold land air common along mountainous coasts of high latitudes. (Webster's VIIth Ed.) Applied in the Aleutians to any sudden storm, usually not forecast, with winds up to 100 knots which wrecked havoc with inadequately secured aircraft, buildings and other facilities.

ZERO — Early WWII nickname for the Zeke, Japanese carrier or land based single seat fighter, Navy Type O, Mitsubishi A6M.

COVER ILLUSTRATION

Bombing Squadron-135 PV-1 **Ventura** flying over the desolate Aleutian Islands. This PV-1, from VB-135's first deployment, features late 1942 paint scheme. The national insignia has been updated in the field to current marking practice by the addition of bars to the insignia stars. "Old dogs" were NEVER repainted in the combat areas, only national insignia and unit markings were kept current. Four stars were applied, at the factory, to the fuselage of the early PV-1s. With the advent of Tri-Color paint scheme, in 1943, the national insignia was applied to the fuselage only under the turret.

U. S. Navy photo, via National Archives, oil painted by Paul R. Matt 1976

Typography
TRADE TYPE
San Gabriel, California 91776

USN via L. A. Patteson

Casco Field, Attu. Home of the Empire Express.

Forbidding landscape of the Kamchatka Peninsula of Siberia seen over the engine of a PV-1 Ventura.

FORMATION AND COMPOSITION

The scene is the bleak, treeless Aleutian Islands, where the Willi-Waws blow and the depressing weather is enough to challenge the sanity of man. Sunny days are rare. Conditions are usually overcast, unpredictable snow and sleet with bitter cold that penetrates to the marrow. Even during the brief summer season, the pinciple factor is fog. The Japanese current is a warm stream of water as it leaves Japan but when it reaches the Aleutians it meets the freezing conditions of the Bering Sea. This moist air mixture over the Aleutian chain causes condensation in the lower levels and a dense fog blanket constantly envelopes the entire area.

It was here in the shrouded Aleutians, during the early part of World War II, that Admiral Yamamoto chose to make a strategic military diversion in an attempt to draw the depleted U.S. Fleet away from the Battle of Midway. Fortunately, Admiral Nimitz was not deceived, and the remnants of the U.S. Fleet met and soundly defeated Yamamoto at Midway. The ultimate outcome of World War II was decided in a brief six minute period that day in June 1942 when SBD *Dauntless'* from the Aircraft Carriers *Enterprise* and *Yorktown* pushed over into their dives. Meanwhile, the American forces in the Aleutians were forced to deal with the Japanese thrust with what they had. PBY *Catalinas* and OS2U *Kingfishers* went out as dive bombers; some PBYs were called upon to make torpedo attacks. They were not intended for such tactics, but they were airplanes when airplanes were needed! In the ensuing confusion the Japanese occupied Kiska and Attu.

USN via M. A. Mason

Wearing 18 bombing mission symbols, a pair of VB-135 PV-1s thunder their way to Paramushiro, 1944.

In 1943, a new star appeared in the Aleutian skies-the Lockheed PV-1 *Ventura.* This hit-and-stun attackbomber was originally produced for the British as a subhunting specialist, but its speed, maneuverability and stamina made it too good to be tied down to patrol work exclusively. The Navy's experience during 1942 with PBY *Catalinas* had demonstrated the vulnerability to enemy fighters of seaplanes operating in combat areas. The U.S. Navy requisitioned the PV-1 from British production and by 1943 had virtually monopolized the *Ventura* production. The PV-1 proved to have superior offensive and defensive capability, which was so desperately needed. It was really a fighter-bomber with a rated top speed of 280 knots (322 mph), its two powerful Pratt & Whitney R-2800-31 engines delivering 2000 horsepower each at sea level. It could maneuver like a fighter and perform beautiful "jinking" evasive action. Lockheed sent a demonstration test pilot to Naval Air Stations along the West Coast who consistently performed an aileron roll on takeoff with one engine feathered! The PV-1 flew well on one engine, but required careful observance of the minimum single-engine speed for adequate control. Its high wing loading of 56 pounds per square foot made landing an experience, but the utilization of Fowler flaps reduced the 91 knot clean stall speed to an acceptable 78 knots (90 mph). Most critical was the minimum single-engine speed. Everyone "sweated-out" takeoffs. Standard Operating Procedure (SOP) was to hold it on the runway until 90 knots was indicated. At that point acceleration was so rapid that single-engine control was practically assured. At 3000 to 4000 pounds over the maximum gross overload, it took all of the 4500 foot runway at Attu to do it, though.

In emergency situations the *Ventura* was capable of more speed than the rated 280 knots. The Pilot's Manual stated NEVER draw more than 52 inches of Manifold Pressure and then only for 5 minutes. All this was quickly forgotten with enemy fighters on your tail. The throttles were "firewalled"-pushed forward until they bent and left there sometimes for as long as forty-five minutes. In the frigid air of the northern Kuriles the results were 56 inches and about 2200 horsepower per engine. One VB-135 pilot reported that after out-running some enemy fighters (*Hamps*) over Paramushiro, he throttled BACK to 54 inches. He had her in high-blower at sea level and had severely abused the engines. Is was necessary to change both engines after his return to Attu, but he made good his escape.

During WW II, a plane going into enemy territory without fighter protection needed one of two things; lots of guns, like the PB4Y-2 *Privateer,* or plenty of speed. With the PV-1 it was speed. In 1943 speeds of 280 knots were considered quite hasty.

One of the principal requirements of a patrol aircraft was good visibility, and in that department the PV-1 did not fare well. The view from the pilot's cockpit was excellent, but there was no good location for lookouts aft, except the turret. An average size man could stand and look out of the astro hatch, but a tall man was uncomfortable. However, the PV's sophisticated radar compensated for any lack of visibility. The ASD-1 Search Radar was the *Ventura's* "Ace in the Hole". The antenna, in the nose, scanned 240° ahead of the plane, and to an experienced operator the ASD gave not only target distance and bearing, but a TV-like picture of topographical features. The first PVs in the Aleutians acted as seeing-eyes for 11th

Air Force bombers. Suddenly weather which had been thought impossible, became a tactical advantage for bombing the Japanese positions on Kiska.

In July 1943, B-24 *Liberators* of the 11th Army Air Force made a successful high altitude raid on Paramushiro. The first raid achieved complete surprise. A second and third raid followed in August and September. They were disasters! Enemy fighters were airborne and waiting, the anti-aircraft fire was devastating. They lost almost half their planes and their enthusiasm for more raids on the Kuriles. The Army decided that daylight raids on the Kuriles were suicide, their operations were needed elsewhere, and replacement aircraft were almost impossible to obtain.

In 1943, after a costly battle, U.S. forces recaptured Attu, westernmost island in the Aleutian chain. The Japanese, while occupying Attu, had attempted, unsuccessfully, to build an airstrip on the northern slopes of the island. While the fighting was still in progress on Attu, the 18th Army Engineers started an airstrip at Massacre Bay, on the east side of the island. By August 1943, Fleet Air Wing Four had moved detachments of PV and PBY squadrons to Attu and Shemya, a small flat island about 25 miles east of Attu, and regular patrols were inaugurated. The PVs began flying anti-aircraft patrols toward the Kuriles, the northernmost of the Japanese islands, which were only about 650 miles from Attu. Not since the Doolittle raid on Tokyo in April 1942, had Japanese soil been within range of American aircraft.

At this point enter Commodore Leslie E. Gehres, Wing Commander of Fleet Air Wing Four, and the *Empire Express* was born. On 16 November 1943, Lt. H. K. Mantius of VB-136 took off from Attu and flew to within 30 miles of the southern tip of Paramushiro. He was airborne nine hours and thirty-five minutes, landing at Attu with 135 gallons of fuel remaining. This was the longest search that had been flown in a PV-1 and Lt. Mantius and his crew were the first Navy crew to see Japanese home territory. This flight had great significance. It demonstrated the relative long range of the PV-1 and that it could be flown on missions to Paramushiro. After that flight, VB-139, VB-135 VB-136, and VPB-131 regularly made the *Empire Express* run, weather and takeoff conditions permitting, until the end of WWII. These four squadrons were continuously based at Attu, on a rotation basis. Two squadrons usually being deployed on Attu at any given time and two reforming with Fleet Air Wing Six.

CINPAC wanted photographic intelligence of what the Japanese were doing in the Kuriles. Commodore Gehres' *Empire Night Express* was initiated by VP-43, a PBY squadron, with Lt. Riddle making the first night flight over the Kuriles on 20 December 1943. By installing special camera equipment in the nose of the PV-1, it was possible to make night photo-recon flights over the northern Kuriles. Special racks were installed in the cabin to carry four magnesium photoflash bombs of one million candle power brilliance. The K19-A camera in the nose, was equipped with an automatic shutter, triggered by a photo-electric cell from the photoflash bombs. The flash bombs were thrown out of the tunnel hatch by a crewman. By carefully timing the release of the bombs in relation to air speed, the proper interval for overlapping picures could be obtained.

A 280 gallon fuel tank was fitted in the aft part of the bomb bay, leaving racks for three 500 pound bombs forward. Cabin fuel tanks and an extra oil tank were also installed. All armor plate aft of the center of gravity and the tunnel guns were removed. In this condition the PVs were dangerously overloaded and suffered considerable tail-heaviness. Lockheed had established 31,000 pounds as the Maximun Gross Overload for the PV-1. With all this added fuel, oil and equipment, *Empire Express* PVs regularly took off at 34,000 pounds, or

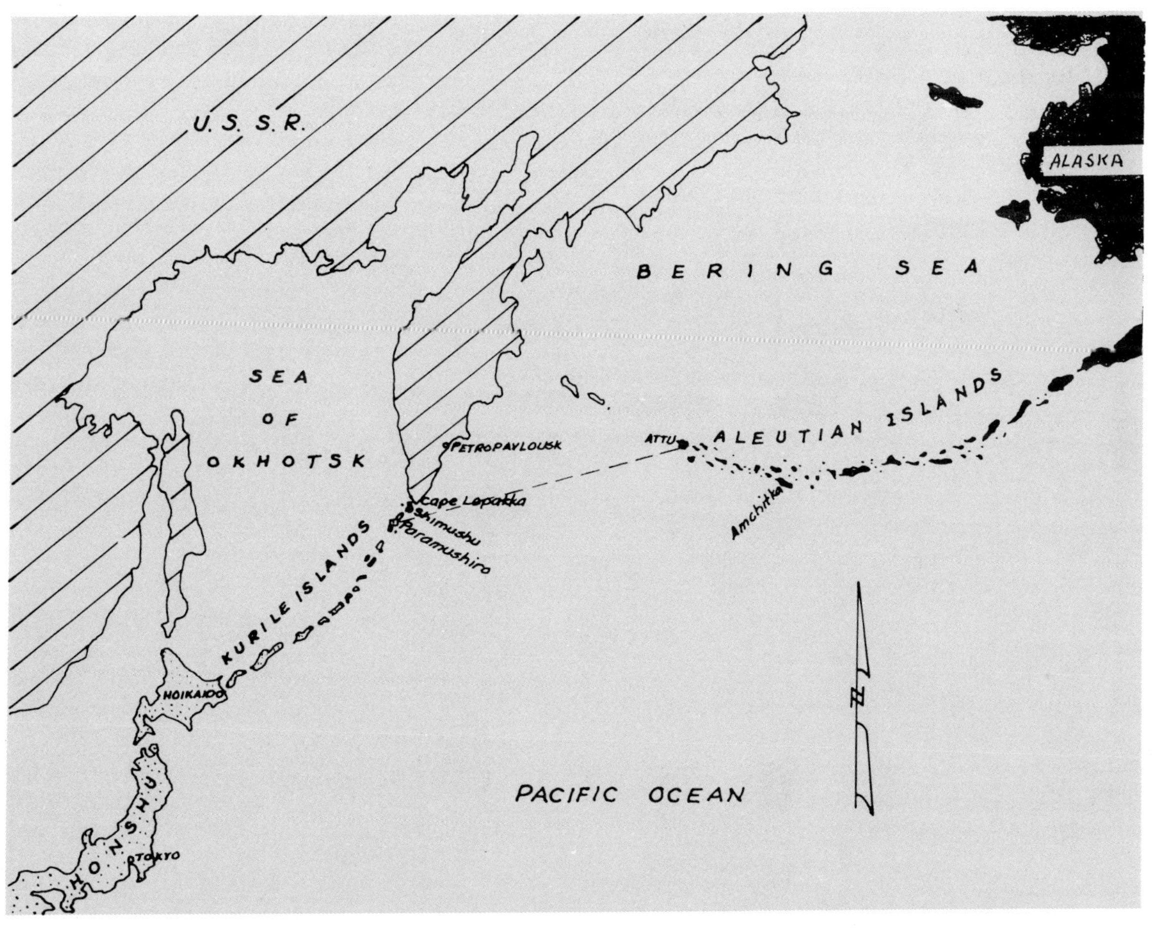

Contrails whip off prop blades as Lt. Al Moorehead takes off from Attu in 1944.

USN via R. R. Larson

3,000 pounds over the manufacturer's Maximum Gross Overload. All this resulted in an extremely dangerous takeoff situation. Several planes and crews were lost on takeoff. The photoflash bombs also constituted a real hazard. One crew was lost in a water landing, after takeoff, when the flash bombs and fuel overload exploded.

The normal PV crew complement was five men, with the co-pilot doubling as the navigator. Due to the extreme distance to the target, and the navigational problems present in the Aleutian area, the *Empire Express* carried a six or seven man crew. An additional Officer navigator, and/or an additional Enlisted navigator, enabled the co-pilot to devote full time to flying.

In flying land-planes across the freezing North Pacific, far beyond the plane's normal range, fuel conservation was of utmost importance. Lockheed recommended a constant cruising speed of 145 knots and the pilot set the power so he could maintain this. In the early part of the flight, power settings were fairly high, but as fuel was consumed and ordnance expended over the target, the power settings would be brought back. The Navy developed a "How-goes-it" curve, which was a graph of time-distance over fuel-remaining. The pilot had to stay ahead of this How-goes-it curve at all times. If he fell behind, there was a point beyond which he could not go and still make it back. VB-135, on their second tour of duty in 1944, had two planes which could not stay ahead of the How-goes-it curve. Different crews flew them on different missions and each time they were forced to turn back before reaching the target. These PV-1s were labled "Gas Burners" and shoved aside.

On one occasion there were not enough planes in an "up" condition for the mission and Lt. J. F. Rumford was assigned one of the "Gas Burners". He reasoned that since he would not make it to the target, he would "shoot the coal to it", and get the flight over with. He cruised at 155 to 160 knots, staying ahead of his How-goes-it curve all the way. When he returned, he had more gas left than anyone. The PV-1 had two angles of attack at which it would fly that heavily loaded. If you flew at the slow speed, you had high angle of attack, and high drag. However, if you gave it more power, bringing the plane over on the other angle of attack, she would really stay up there. So, quite by accident, it was discovered that a higher cruising speed gave better gas consumption.

The engines were started and warmed-up in revetments. After taxiing to the end of the runway where a fuel truck waited, the engines were killed and the tanks topped-off, thus takeoffs were made with every possible drop of fuel aboard. These operations assumed a carrier deck atmosphere, with the duty officer going from plane to plane giving last minute verbal instructions. No radio was permitted, even on takeoff. Radio silence was observed all the way to the target, everyone flying on instruments, every crew on its own until the target was reached and an attack made . . . then the radios would come alive with chatter.

A large 18-inch diameter loop antenna was installed under the nose. This loop was needed to combat precipitation static, which was a serious problem in the area. A battery powered Lear portable radio receiver was issued to each crew. These units covered not only the broadcast band but also the low frequency aircraft bands. One of VB-135's crews was saved by the Lear portable. They lost their electrical system and were able to get down using the Lear hooked up to the main antenna.

Weather was the number one operational problem. The fog was usually so dense that the flyable condition was anytime in which the wind was coming from the opposite side of the island. With the wind in that direction, the air was lifted over the mountains. The moisture was wrung out by the lifting and some of it would be precipitated out in the mountains. The compression-heating on the lee side of the mountain would be sufficient to form a bubble, and that is where the airstrip was located. The Weather man's chief problem was being able to predict that a bubble would form. The missions were usually of 9 to 10 hours duration, so he had to project whether the wind would be coming from the proper direction for the field to be open. The only reports he could depend on were those received from the Guard Ship. This ship, a small AVP, was deployed about 200 to 250 miles out, toward the target. They sent frequent weather reports, the one which was of principal interest was the wind direction. Attempts were made albeit not very successfully, to pick up Russian weather reports from Kamchatka, particularly the weather at Cape Lopatka. Usually these were garbled. Also they were many hours old, so the weather man really did not have much to go on. There were occasions when the bubble was collapsing as the planes were making their final approach. If this happened, the only alternate field was Shemya, some 25 miles from Attu. Shemya was a flat island so if Attu was closed in, chances were that Shemya had been closed in for some time. There really was no alternate, except ditching in the freezing water. In such an event, anything in the way of a rescue had to be extremely prompt. Even in a rubber raft, if you got wet, and there was no way of getting into the raft without getting wet, you could only last about 20 minutes before you died of exposure.

The Aleutian PV squadrons were 18 crew squadrons, with 15 PV-1s assigned, and 12 active planes at all times. Fleet Air Wing Six furnished replacement crews and aircraft to compensate for losses as they occurred, keeping the squadrons up to full strength. During VB-135's second deployment, they lost the replaceemnt crew for a replacement crew! The squadron organization was streamlined and composed almost entirely of flight personnel. A typical squadron consisted of 54 flying officers, a Personnel Officer, an ACI officer, 72 flying aircrewmen (AMMs, AOMs and ARMs), 2 Chief Petty officers (Leading Chief and Line Chief, both aviation ratings), and one or two Yoemen. The Commanding Officer, Executive Officer, and Operations Officer were all flying Patrol Plane Commanders.

The Chiefs were responsible for keeping the planes in good repair and seeing that they were properly loaded for each flight. The maintenance, refueling and arming duties were performed by HEDRON (Headquarters Squadron). This was a separate command of the Fleet Air Wing. The Personnel Officer was responsible for all records, reports and general supervision over the Administrative Department. The Yoemen was in his domain.

Having more crews than airplanes, 12 crews had a plane assigned to them, which they shared with two other crews. In this way each crew would fly one particular airplane, or one of two airplanes. This practice was a tremendous morale booster and the crews lavished care on them. Many were decorated with beautiful examples of L'Art de Femme.

On 11 June 1944, Lt. J. P. Vivian, on a daylight weather hop, found weather conditions right and continued over Shimushu where he found a new airfield full of Mitsubishi *Bettys*. He was able to get some good pictures with a K-20 camera. The finished prints definitely located the field. The *Bettys* were equipped as Torpedo Bombers and since there was an American task force in the area, it was imperative that the field be neutralized. Lt. M. A. Mason, Exec. of VB-135, and five other volunteer crews took off on 12 June in the first daylight raid since the disastrous Army raids in Sept. '43. The mission was a complete success and exploded the myth that it was suicide to fly over the Kuriles in daylight. The raids between 11 and 14 June 1944 demonstrated that daylight missions were much more satisfactory. From then until the end of the war, with few exceptions, daylight intruder missions were carried out. In these operations, low altitude attacks were made with planes striking in pairs and fours. These raids were most successful when low cloud cover prevented enemy fighters from making overhead runs. Bombing was much more accurate from such low levels. The extreme distance to the target precluded the feasibility of fighter escort. Most fighters of the era would have been hard-pressed to keep up with the PVs, had they possessed the required range.

In the daylight missions, the approach to the target was made "on the deck". The props were almost chewing the water; the slipstreams would stir up the salty water enough to leave wakes that could be seen behind each plane all the way to the horizon. This kept the PVs under enemy radar until the last possible moment. The raids were sporadic and the enemy's first warning to scramble came not much before the PVs streaked across the beaches. Even among the best trained, it takes a few minutes to communicate the warning and have fighters airborne.

Due to the inherent difficulties of night photo-reconnaissance, small coverage per picture, and few available shots in four flash-bombs (most of the pictures were of clouds), the night missions did not prove successful.

Crew of Lt. "Butch" Mason manning 3V3 in preparation for a daylight mission of 12 June '44. Scattered quonset huts dot the hillside in the background.

USN via M. A. Mason

U.S.N.

Lt. L. A. "Pat" Patteson taking off from Attu on a daylight photo mission of 14 June '44. The PV-1 was specially rigged with a 480 gallon fuel tank which filled the bomb bay, five cameras in the nose, and carried an extra crewman — a Photographer's Mate. Crew threatened the PhoM with lynching if the cameras jammed. Fortunately several hundred excellent pictures resulted.

The Japanese now feared an invasion from the Aleutians; frantically building up their defenses in the Kuriles. Fighters were encountered more frequently but the PVs proved able to deal with them. Several were shot down in head-on attacks with the bow guns and the turret gunners began to run up an impressive score. If the fighters were able to chase them, the situation was usually in the PV's favor. Both the *Venturas* and the fighters were gulping fuel at the maximum rate but there was a big difference in what each was accomplishing. The PVs were heading for home; the fighters were heading away from their base. With an eye on their fuel gauges, the enemy usually followed for awhile, staying well out of turret range, finally did an Immelmann turn and went back to their home field. If they were able to close on the PVs, they were prevented from making the typical fighter pass due to the PV's speed. They were nearly always forced into making stern chases. There was no coming up from underneath, because of the PVs evasive tactics of hugging the water. The closure rate being so slow, the fighters presented a simple zero deflection shot to the PV gunners. They soon learned that the turret's torrent of accurate .50 cal. fire was deadly. The enemy became less and less aggressive in pressing their attacks.

After June 1944 the *Empire Express* changed from photo-recon to intruder operations. VB-135 was relieved by VPB-131 in October 1944 and they brought a new weapon to the *Empire Express* — aircraft rockets. The PV-1 being a fighter-bomber, 131 went at it in that fashion. High-speed strafing runs included rocket attacks. VPB-131's PVs carried the "Chin Gun" package, increasing their forward fire power from two to five guns. No bombs were carried in this configuration. The bomb bay was filled with a 480 gallon fuel tank, which greatly extended their loiter time over the target area. The squadrons attacked targets of opportunity—canneries and fishing vessels ranking high on the list. The strategy being to hit Japan in the "Bread Basket" (or Fish Basket). *Empire Express* operations were geared to coincide with the Navy's operations in the South Pacific and were stepped up in intensity prior to any big push in that area. Throughout their long operations in the Aleutians never once did the PV-1 *Venturas* have fighter escort. As a matter of fact, on several occasions they were called upon to fill the escort role themselves for Army heavy bombers.

John Daws

Lt.(j.g.) Stafford B. Mantz clowns at the plight he lived with almost daily. The laundry tie-lines (rope) belt was non-regulation.

USN via L. A. Patteson

Refueling one of VPB-139s Harpoons on Attu.

When VPB-139 arrived on Attu, in March 1945, a new PV appeared in the Aleutian theatre, the PV-2 *Harpoon*. Conceived early in WW II to replace the PV-1 and to correct some of its faults, the *Harpoon* more accurately reflected the patrol-bomber concept. The early PV-2s were plagued with teething problems and arrived in the Aleutians before many of these discrepancies were resolved. Although range was increased and flying characteristics improved, the top speed was reduced. It is doubtful if the Japanese ever recognized the difference or realized this reduction in performance. By the time the Harpoon's arrived, Japanese fighter pilots had developed a keen respect for anything that resembled a PV!

Why the PVs emerged with so little recognition in WWII histories remains a mystery. They lived down their early reputation as bad actors and the Navy found them to be safe and reliable airplanes. All who flew them grew to love and appreciate the little beasts. All the crews who had the misfortune of receiving battle damage, or experienced mechanical malfunctions over the Kuriles and had to make a forced landing, or parachuted into Kamchatka, received rather ungrateful treatment from Russian allies. Some of the men of the *Empire Express,* who had to land in Russian, stated later that if they had it to do again they would have risked a ditching in the freezing Pacific waters. Some were fired on as they made final approaches to Petropavlosk, with flaps and landing gear down! Although they endured many hardships, all were eventually returned to the United States, usually through Europe.

In retrospect wars are viewed differently by the participants. One weighs the loss of friends against the victory to which they contributed everything. Information, which became available after the war, indicated that the Japanese took the bait dangled by the *Empire Express* and the losses were, indeed, justied. Estimates vary from 1/6 to 1/4 of the Japanese air forces were diverted to deal with an anticipated invasion from the Aleutians. These were men and machines which the real thrust from the South did not have to encounter. This caused great frustration by the Japanese High Command who so badly needed these resources in the South Pacific. The *Empire Express* was undoubtedly a factor which effected the ultimate outcome of WW II in the Pacific.

In 1972, the "Airdales" of Fleet Air Wing Four held a reunion at Oak Harbor, Washington, moderated by Admiral James S. Russell, USN (Ret.). One hundred sixty-five veterans of the Aleutian squadrons attended. Also attending was Vice Admiral Hiroichi Samejima of the Japanese Navy, who, as a Lieutenant in 1942, had led the Japanese planes attacking Dutch Harbor. Admiral (then LCdr.) Russell was the Skipper of VP-42 at Dutch Harbor in 1942. VAdm. Samejima attended at the invitation of Adm. Russell. The two first met in 1955 and became good friends. The wounds of war eventually heal.

PV-2 Harpoon of VB 139. Their PV-2s were delivered in the 1944 tri-color scheme; Dark Sea Blue top surfaces, graduating to intermediate blue side surfaces and off-white under surfaces.

U.S. Navy

EMPIRE EXPRESS SQUADRONS DEPLOYED ON ATTU
10 AUGUST 1943 to 16 AUGUST 1945

On 1 Oct. 1944, the designation of all PV squadrons was changed from Bombing (VB) to Patrol Bombing (VPB)

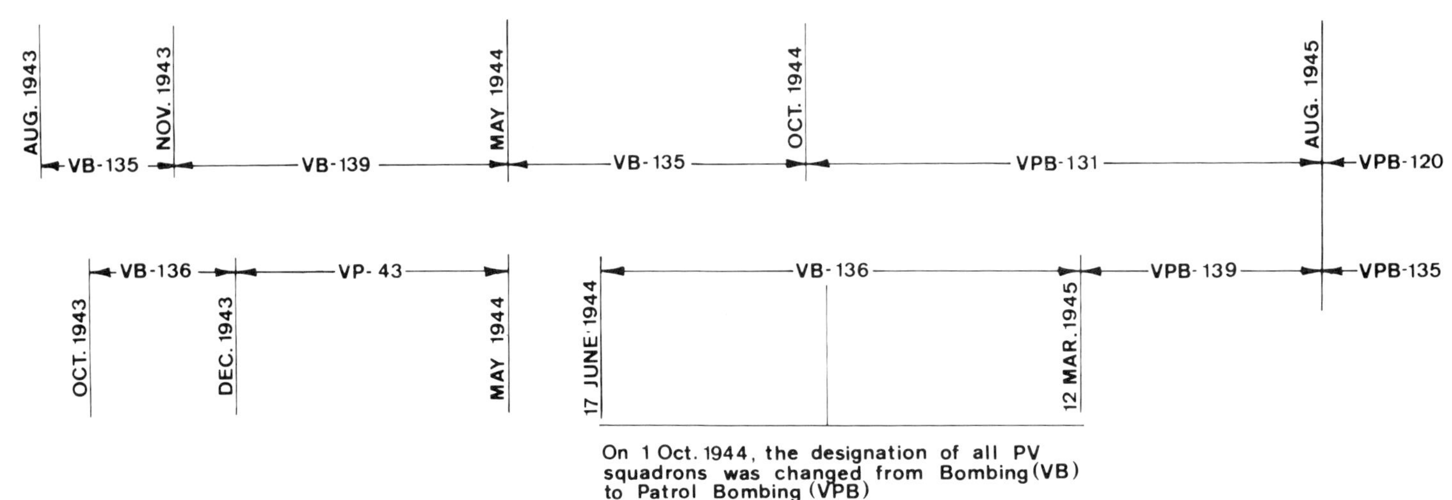

Typical sector search carried out by the Empire Express.

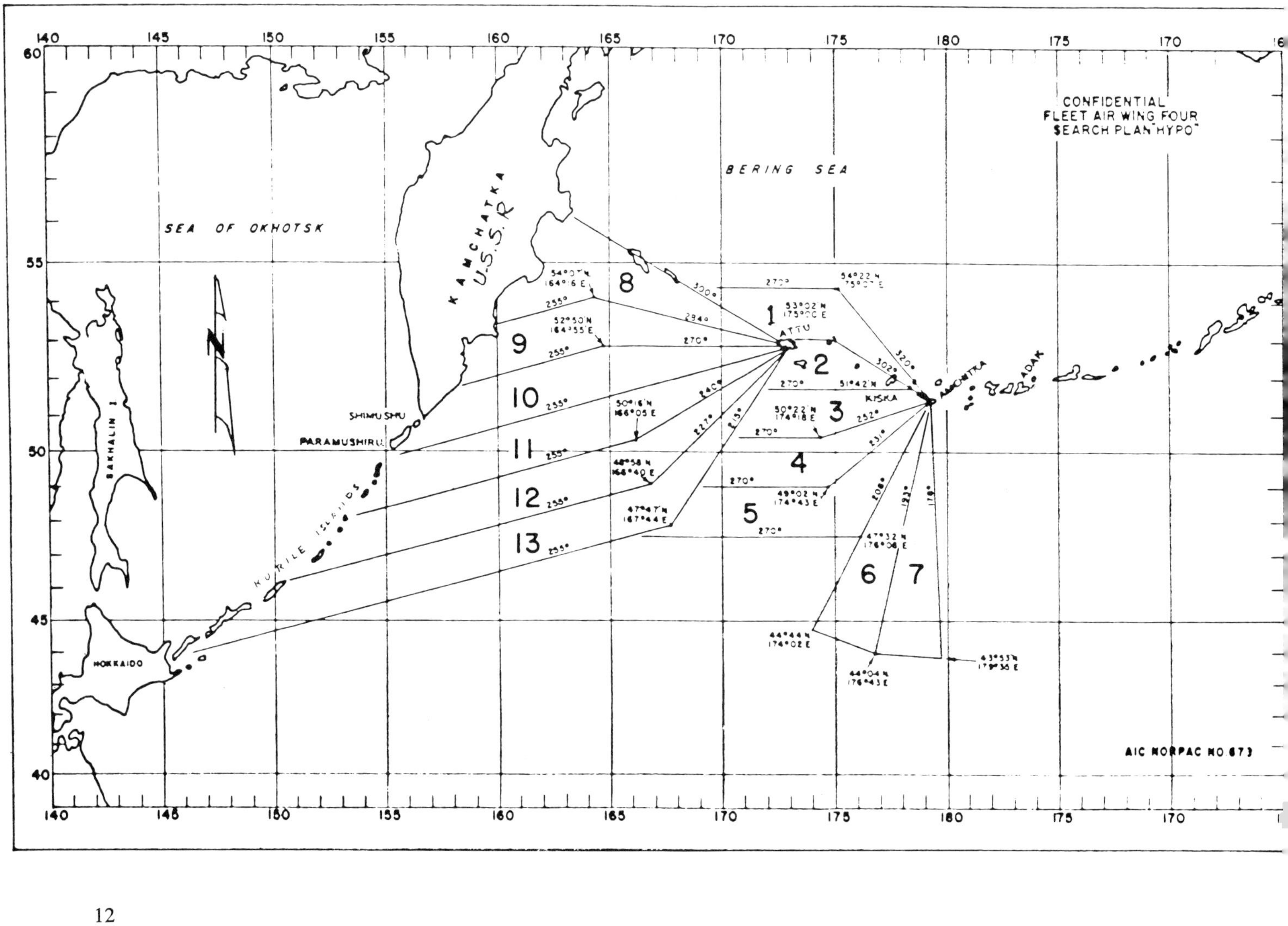

History of Bombing Squadron - 135

VB-135

Bombing Squadron 135 was commissioned 15 February 1943 at Ault Field, Whidbey Island, Washington, under the command of LCdr. P. C. Williams, USNR. The nucleus of the squadron was obtained from VP-42. This squadron, flying PBY *Catalinas,* has been operating in the Aleutians and had made an impressive record in the defense of Dutch Harbor when it was attacked by a Japanese task force in June 1942.

The squadron was the initial Navy PV-1 *Ventura* squadron to operate in the Pacific, and the Aleutian area. The transition from PBYs to a "Hot" airplane like the PV-1 was not easy. The ex-PBY pilots, accustomed to the slow, docile P-Boats, found the powerful PV-1 skittish. Slowly they gained confidence and respect for them however, despite the loss of a crew in a training accident. The squadron trained at Whidbey Island from 15 February to 25 March 1943 and whatever the training syllabus, the time allowed for training was inadequate. In the final analysis this left much to be desired. The first pilots completed training with from 60 to 80 hours time in the PV-1 type aircraft. Instrument time consisted of work under the hood, which was not suitable for flying in the Aleutian area. Being P-Boat pilots, they kept the "keel" next to the water and would never let go, no matter what the weather conditions were.

Upon the completion of this all too brief training period, it befell VB-135 to pioneer cold weather patrol with an airplane which had originally been designed as a fast attack-bomber. They were assigned to an area which consisted of a narrow chain of volcanic islands, where the most treacherous weather can be found anywhere in the world.

On 25 March 1943, the squadron departed Whidbey Island for Adak. Due to bad weather, it took them three weeks enroute, not arriving until 12 April 1943. At Adak the squadron flew photo-reconnaissance missions over the Japanese held islands. Special equipment was scarce, and the planes were not equipped with proper camera installations. Ingenuity persisted though and fair results were obtained with hand held K-20 cameras.

On 5 May 1943, the squadron set down on Amchitka and began flying late afternoon patrols. Ten enlisted navigators soon joined the squadron, increasing the crew strength to six men. Amchitka was within bombing range of Kiska and it was this fact, plus the faith LCdr. Williams had in the merits of radar bombing, that led to the successful raids on Japanese held Kiska by Army planes. These notable missions were led by PV-1 *Venturas* of VB-135, which were equipped with the more advanced ASD-1 radar. Although the results obtained by this type of bombing were completely satisfactory, VB-135's primary mission remained Patrol. The reconquest of Attu began in May 1943 and it was feared that a strong Japanese task force would try to intercede. The PV's were loaded with torpedoes but the weather closed in and the attacks were scrubbed.

During the squadron's stay at Amchitka, two of the personnel were killed. The accident occurred during takeoff, when the pilot, thinking he was airborne, retracted the landing gear. The plane settled back to the runway and crashed. Quick action on the part of a nearby Army B-24 crew saved the lives of three members of the PV-1 crew.

Living conditions at Amchitka were primitive. The personnel were housed in tents. Food consisted of Army field rations. The runways were rough and it was a standing joke among the Navy men that the Army engineers would not build a runway unless it was 90° to the prevailing wind. In their criticism however, they overlooked the fact that there was no prevailing wind. During the last 17 days of the squadron's stay at Amchitka, the weather closed in and nothing moved. It was during this period that the Japanese evacuated Kiska.

A destroyer patrolling the entrance to Kiska Harbor claimed two direct hits on a Japanese submarine, firing with the aid of radar. Lt. (jg) Davidson of VB-135 photographed a sub near Twin Rocks . . . broken up on the shore. This was sufficient evidence to credit the destroyer with a definite kill. When the ship put into Attu in August, they contacted Lt. Davidson and made him a present of a box of food plus wining and dining him aboard ship.

The occupation of Attu had been completed and the squadron set down on a partially completed air strip at Alexai Point on 10 August 1943. The same conditions prevailed as at Amchitka, cross winds on a rough runway, living in tents, field rations amid a generally miserable existence.

Lt. (jg) L. W. Fischer had flown to Adak for maintenance work on his plane. Just before returning, he purchased twenty cases of beer and loaded them in the bomb bay. Weather temporarily grounded him at Amchitka. That night HEDRON pulled a routine hydraulic check on his plane. When they opened the bomb bay doors, out dropped ten cases of beer — a rare and welcome sight to any maintenance crew. Ten cases were immediately salvaged but Lt. Fischer arrived back on Attu with only half rations remaining. It took resourcefulness to acquire such luxuries but even so, it had to be shared with shipmates.

PV-1 Venturas of VB-135 on their first tour of duty in 1943.

USN via L. A. Patteson

Kakumabetsu airfield on Paramushiro, taken by photoflash bomb at night by VB-135.

U.S. Navy

From Attu the squadron flew searches 500 miles west and south of Attu. Due to Japanese air patrols from Paramushiro, the squadron was called upon to fly anti-aircraft patrols toward the Kuriles. Results of all these patrols were negative, however.

On November 1943 the squadron was relieved of duty and departed Attu for Whidbey Island, returning in four planes. The remainder were unserviceable. This group had been the PV-1 pioneers in the Aleutians. They had sweated out a tour of duty where the weather constituted an extremely great hazard. What they lacked in experience, they made up in courage.

VB-135 was officially reformed on 3 January 1944, under the temporary command of Lt. M. A. "Butch" Mason, who later became the executive office of the squadron. "Butch" came to VB-135 from Pensacola, where he had been an instrument instructor in multi-engine landplanes. He learned instrument flying at Pensacola under Jack Thornburg, formerly chief pilot for TWA and an instrument pilot of great renown. Because of Lt. Mason's background, he was able to train the squadron in this type of flying. Fleet Air Wing Six subsequently furnished four PV-1s. The squadron was divided into four groups and flew around-the-clock in shifts. HEDRON FAW-6 managed to keep three of the four planes in an "up" condition at all times. They flew the FAW-6 syllabus regardless of weather, day or night, rain or snow. The only breaks in the strenuous program were special exercises such as torpedo drops.

In February 1944, Commodore Gehres flew from Adak to Whidbey to brief the squadron on his plans for them. They were to be a special unit, whose mission would be to photograph the Japanese home islands of Paramushiro and Shimushu at night. The training stressed navigation, low level bombing, radar bombing, instrument and night flying. Training in night photography with photoflash bombs was included. During the squadron's first tour of duty in the Aleutians, the need for more accurate navigation came to light. As a result, eighteen officer nagivators and an equal number of enlisted navigators were assigned to the squadron. Eventually there were eighteen full crews, each consisting of three officers and four enlisted men.

VB-139 was engaged in the *Empire Express* runs at the time Commodore Gehres instituted his photo-recon operation. VB-135 began receiving its own PV-1 aircraft early in February and work commenced on the installation of the special equipment. The planes were equipped with bow cameras, mounted in the flat glass panel under the nose. These were Fairchild K19-As, with a 13½ inch focal length lens for night work. The camera was operated electrically and tripped by a solenoid. Racks for four photoflash bombs of one million candle power were installed in the cabin. The aft portion of the bomb bay was fitted with a 280 gallon fuel tank, leaving space for three 500 pound bombs in the forward racks. Additional racks were fitted inside the fuselage for a dozen 26 pound fragmentation bombs. These were to be used for harrassing purposes and located under the turret, to be thrown out during the photographic runs. External drop tanks were also carried, since the round trip to the target area (1500 to 1800 nautical miles) would be greater than the normal range of the PV-1. Because of the additional gas and equipment, all armor plate aft of the center of gravity, and the tunnel guns were removed. Even with this, the gross overload limits of the PV-1 was strained with an extra 2500 to 3500 pounds. These load and balance problems made night takeoffs frightening.

Late in the training period, LCdr. P. L. Stahl assumed command of the squadron and Lt. Mason became executive officer. On 16 April 1944, the squadron departed Whidbey Island for Yakutat and then flew up the chain of islands to Adak, arriving there on 21 April 1944. The squadron immediately reported to Headquarters, Fleet Air Wing Four only to learn that they were scheduled for four days of intensive training in the operation of LORAN. This was a new thing and very hush-hush. LORAN is a radar-radio navigational system using medium or low frequencies enabling an aircraft to position itself by finding interrelated pulses transmitted by pairs of ground stations. It is similar to an earlier radio "Gee" system but with greater range and far more accuracy. Almost immediately upon arriving, HEDRON began installing the units in the PVs. After a number of familiarization flights getting used to the electronic marvel, VB-135 left Adak for their new base of operations, Casco Field, Attu, arriving there on 4 May 1944.

The primary mission of the squadron was to fly night photo-recon missions over the Northern Kuriles in an attempt to discover the full extent of Japanese military activities. These flights were the outgrowth of their predecessor, VB-139, who had proven that by taxiing the PV to its greatest capacity Paramushiro could be bombed in nuisance raids. The crews never took off in anything less than a 34,000 pounds gross weight condition.

On 5 May 1944, the squadron initiated operations with a nine-plane strike against Shimushu Island. Night reconnaissance and bombing missions of four to eight planes were carried out every night the weather permitted, until 12 June 1944. Unfortunately, three planes and crews were lost in these operations.

The squadron soon became quite proficient in night radar bombing. On one occasion, Lts. Mason, Patteson and Sparks made a night bombing attack on an airfield on Shimushu. They attacked from three different directions, using three different radar aiming points. The PVs were armed with three incendiary clusters, about 200 bomblets assembled in one big

USN via M. A. Mason

Crews manning their planes in revetment, Casco Field, Attu. After briefing, crews were taken to their planes by the big gray bus.

U.S.N.

VB-135 planes in the revetment. Black 2 is squadron number, 889 is the last three digits of the BuNo. 48889.

cluster and hung on a 500 pound rack. Each of these incendiary-clusters would paint a line on the ground 50 yards wide by 150 yards long. If everything went right, the bombs being dropped by an Intervalometer, the three bombs would form an extended strip some 450 yards long. On this particular night, strike photos revealed that all three strips actually crossed, forming a star burst pattern across the target. Excellent results were obtained from this technique of bombing, including a number of secondary fires among *Betty* bombers on the field.

On the photographic missions team work was essential, for the flash bombs had to be thrown out of the tunnel hatch at the proper intervals to get over-lapping stereo pictures. The bomb run itself was made by radar at an altitude of 8,000 to 12,000 feet and at a ground speed of 210 knots (242 mph). The flash bombs were set to go off at 4,000 feet and this flash tripped the camera shutter.

A new airfield had been reported built on Shimushu, on which possible long range Japanese *Bettys* were based. Lt. (jg) Blakeney volunteered to make a daylight run in an attempt to find this field. Unfavorable weather was encountered over the target area and he had to return. The photos taken during the flight revealed only mountain tops. On 11 June, Lt. J. P. Vivian, on a weather hop, found conditions more favorable and extended his flight over Shimushu, where he was able to obtain good pictures of the new airfield.

A U.S. Navy Task Force was going to bombard the Shimushu coast and it was most advantageous to prevent Japanese bombers from spoiling the show if at all possible. Lt. Mason, the executive officer, was quickly ordered to lead a six plane strike force in an attempt to neutralize the enemy airfield. He was told to find volunteer crews for the five other planes that would make up the strike. Lt. Mason approached the leading chief, told him his orders and asked him to canvas the enlisted aircrewmen for the volunteer crews. There was no problem obtaining crews among the officers but when Lt. Mason returned to the chief he discovered that he was unable to find five volunteer crews among the enlisted men. Lt. Mason asked why. The chief replied, "Well sir, no one wants to fly with anyone except *his own* patrol plane commander. 135 has the last damn PPCs in the Navy, as far as the eighteen crews are concerned. They will go *anywhere* with *their* PPC, but not with anyone else." As soon as Lt. Mason gave the chief the list of PPCs who had volunteered, their respective crews immediately volunteered also.

The next day the squadron was off on their mission. Due to clouds and fog over the target, radar was used to direct the bombing. Although damage could not be determined visually, they felt certain that the strike was a success. However, Lts. Mason and Patteson swung south and proceeded to survey the west coast of Paramushiro. Breaking into the clear they discovered still another new airfield at Kakumabetsu. It was well under construction, the runways were paved and revetments almost finished. The two PVs banked swiftly and made several strafing runs. Lt. Mason's navigator, Ensign Richard J. Hanlon was able to get some good intelligence photos. While the AMM sat on his legs, Hanlon hung out of the tunnel hatch and snapped pictures with a hand held K-20 camera. When the crews returned home they were exhausted but highly elated over an exceptionally good day's work.

On the 14th, Lt. Patteson, with a specially rigged photographic PV, made a daylight photo run along the full length of both Shimushu and Paramushiro, getting excellent pictures of five major airfields. At the same time, to create a diversion, Lts. Sparks, Vivian, Clapham, Mabus, Bone, and Schutte carried out a raid against Miyoshino airfield on Shimushu. Lt. Patteson described the mission, "By the time we reached the northern end of Shimushu the activity below was a most confused and active encounter. The diversionary force's surprise was successful, but we were all surprised at the beehive they had stirred up. Aboard was my regular crew plus a PhoM to set and activate the cameras. It was CAVU — not a cloud in the sky. At 11,000 feet, we headed for the first

Karabu Zaki on the southern tip of Paramushiro. Taken on the daylight photo run by Lt. Patteson's crew, 14 June '44.

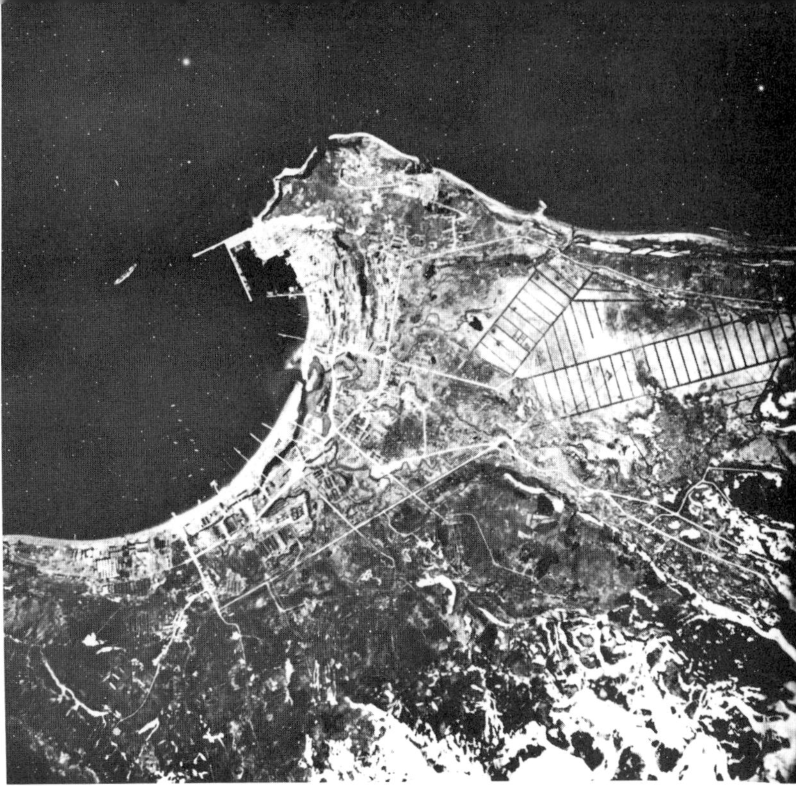

Kataoka on Paramushiro as seen on 14 June '44 photo mission.

target, Kataoka. We had a bomb bay full of extra fuel and I set power for maximum cruise, so we had good speed. Everyone craned their necks for bogeys, but none appeared. All five cameras were turned on as we approached the first field, then the second target was spotted and the crew shouted that they could see planes taking off below. We continued south over Kashawabara, when two bogeys appeared ahead of us and went by before they knew who we were.

"Our next target, Dago Zaki, was about 40 miles to the south. The *Zeros* turned around and gradually began overtaking us. At first they just flew parallel to us, staying a respectful distance from our top turret guns. When they got in close, the turret gunner, Jacobsen, opened fire. Having a very healthy respect for the twin .50s, they quickly jinked out of range. After a few minutes of this, they climbed above us to make the typical fighter passes from 4 o'clock and 8 o'clock high. My navigator stood in the astro hatch, with an intercom mike, and warned me when one began his run. All I had to do was turn into the direction of the fighter's pass and this shortened the angle so steeply that their fire always went behind us. The fighters were faster and more maneuverable at this altitude than we were, but our relative speed was the big equalizing factor. By the time we reached Dago Zaki, we were untouched and felt we had some kind of edge. Of course, during this time we had only caused damaged to their ego.

"By now we were approaching our final target, Dago Zaki, and saw many fighters coming at us from ahead and below. They had apparently come up from the big base at Kurabu Zaki. Much shouting and shooting ensued. The turret gunner, Jacobsen, got a "probable" here — he certainly had plenty of targets to shoot at. Two things were running out; the film magazines and our cool. We managed to get a good oblique shot of Kurabu Zaki. Miraculously we still had no damage. We shut off all cameras and I was only intent upon shaking loose. I made a shallow dive from about 8,000 feet to the deck. I 'firewalled' everything and prevented any of them

All Photos USN via L. A. Patteson

Kashiwabara airfield on Paramushiro, 14 June '44.

Adjusting parachute harness prior to manning planes. PVs were equipped with chest pack parachutes and only the harness was worn in flight. The chutes were stored in the cabin.

USN via M. A. Mason

from overtaking us — they just did not have enough overtaking speed. One or two of them stayed with us for several minutes but Jacobsen had little trouble keeping them off our tail."

Of the six planes in the diversionary force, two were forced to land in Russia with battle damage and the four planes that returned needed engine changes from excessive use of power. Two received damage from enemy AA. On 25 June 1944, Lt. Patteson made another daylight run over Kakumabetsu, showing the development of the field in eleven days. He was attacked by 11 fighters and returned with two bullet holes in the tail surfaces. His turret gunner, Floyd Jacobsen AOM2c, was credited with one 'kill' and one 'probable'.

"Butch" Mason described one of his most memorable intruder missions on 21 July 1944, "After a night takeoff we arrived at the target area alone, just as the sun was coming up. It was one of those rare days when there was not a cloud in the sky. Our orders were to never go over the target alone without clouds. However, it was just daylight and I could see the rotating beacon over the airfield on Shimushu. I figured that no one was up and flying and I had caught them before breakfast. I passed the word to the crew to keep their eyes open for fighters. We headed on in, and just as we got over the airfield, the turret gunner yelled 'Fighters' over the intercom. I will never forget that sight. I leaned forward and looked up. There was a formation of eight *Tojos*. The first one was peeling-off to make a run on us. I must have used up a year's supply of adrenalin right there! We were directly over the field, and just out of instinct, I flew the PV through a split-S and wound up heading straight down at the field. I opened the bomb bay doors, pulled her up a little and pickled-off the bombs. About that time a whole bunch of tracers went by on my left. Several bursts of heavy AA exploded behind us. I remember thinking it strange that they would shoot at us while their fighters were making a pass. The fighters must have flown right through their own flak. I closed the bomb bay doors and headed for the deck. I had full allowable power on and by the time we got to the water we were indicating 310 knots (357 mph). I put the belly right down on the water and at full speed a fighter was just asking for it to make a run on a PV. Only one of the *Tojos* was able to follow us through the split-S and sure enough, he tried a run, but the turret gunner, Richard McGee AOM2c, opened up on him and he broke off his attack at that point. So now we were headed for Attu at 300 knots. I throttled back and everything calmed down."

On 23 July, Lt. Vivian on a CAVU day, attacked and sunk an armed picket boat. Due to battle damage he was forced to land in Russia. On the same day Lt. J. W. Pool was attacked by eight *Tojo* type fighters. He shot one of them down with the bow guns.

On 12 August, Lts. Mason, Rumford, Sparks and Mabus placed three 500 pound bombs directly on the Kakumabetsu airstrip and obtained photos from 200 feet showing his hits.

On 12 August, Lts. Mason, Rumford, Sparks and Mabus bombed and strafed targets on Arido To. Jim Rumford had hit a cannery building with an incendiary and had a good fire going when "Butch" Mason made a low pass and laid a 500 pound GP in the front door. Jim was in position to watch the bomb go in the door. When the delayed action fuse set the bomb off, the building literally blew apart. He came on the VHF radio and blared to "Butch", "Hey, watch that, you blew my fire out."

Lt. Mason was intrigued by the "juicy" target of docks and staging area which he saw and photographed on 12 June near Kakumabetsu. He and the squadron intelligence officer had been studying the photographs when the ACI officer pointed out that all the heavy AA batteries were up on the bluffs overlooking the docks. It was also noted that the guns would not depress below the horizontal.

To take advantage of this they reasoned that the best strategy would be to come in low from seaward, below the cliffs and avoid the AA batteries. On 7 September, Lts. Mason and Sparks on a daylight intruder mission to Paramushiro, found their primary target obscured by overcast. Using radar, they crossed over the mountains of Paramushiro. Mason spotted a hole in the overcast and dived through it, coming out directly over Kakumbetsu. The ceiling was about 1500 feet with good visibility. As he circled to seaward he spotted a small vessel offshore and gave it three strafing passes. This was probably a mistake as it allowed the enemy ashore plenty of time to get ready.

Following the planned strategy he came in below the cliffs, flying right down the docks. He pickled off all three 500 lb. GPs (with delayed fuses) laying them right on the targets. At the end of the docks was something which they missed when reviewing the photos, a 20mm gun position. The 20s raked the PV from bow to stern. Lt. Mason was not aware that they had been hit. He remembers hearing a popping noise: then noticed that the glass was broken in several instruments and smoke began pouring into the cockpit from the nose. He looked across at the co-pilot, Hanlon, who had a gash across his temple and was bleeding profusely. He turned to Tiner, the ARM, who was directly behind him. His face was a mass of blood. Mason grabbed Hanlon by the neck and held the pressure point to stop the bleeding. Hanlon soon realized what he was doing and held the pressure point himself. He unbuckled, got out of his seat and started aft. As he departed Mason yelled, "Get someone up here with a fire bottle". Horstman, the enlisted navigator, appeared with a fire extinguisher and dived into the nose area. The camera installation and oxygen bottles were burning but he soon had the fire out.

The airplane was still flying. Mason had given her full power, climbed up out of the overcast and headed for Attu. The other crewmen attended to Hanlon and Tiner's wounds as best they could. Tiner had sustained the most serious injuries. A piece of shrapnel caught him just above the eye, causing hemorrhaging in the eye ball. Lt. Mason remembers eating his lunch on the flight back and there was blood on the balogna sandwiches (in the collaquialism of the Navy, balogna sandwiches had a different unprintable name) which made it all the less palatable. An ambulance was waiting when they landed. Hanlon and Tiner were loaded in the ambulance. Both men recovered and were able to fly back to the United States with the crew later.

Beginning 15 October and for three solid days thereafter, the squadron provided cover for the Task Force going in to bombard Matsua. Finally on 23 October 1944, the squadron was relieved of duty in the Aleutians by VPB-131. 135 had sustained more losses than any *Empire* Express squadron . . . twelve planes and ten crews. Five of the ten crews landed in Russian territory. During their tour of duty they had flown 287 combat sorties. Although they operated under strict orders not to attack the fighters, only in self-defense, they were credited with downing eight Japanese aircraft. The squadron ace was Floyd B. Jacobsen AOM2c, who was credited with three *Hamps*. Upon arriving in the United States the entire squadron was given thirty days leave.

On 5 December 1944, the squadron was officially reformed with Lt. Mason assuming command. The new squadron was composed of 26 of the old officers and 22 of the old enlisted complement. During the period 1 March to 1 June 1945, the squadron was engaged in training at Whidbey Island. The syllabus consisted of glide bombing, low altitude radar bombing, rocket firing, torpedo drops, aerial mining, sleeve firing, day and night overwater navigation, night radar searches and flare exercises. Seven PV-1 *Venturas* were assigned but normally only three were continuously in operation. Late in May the squadron began receiving their assigned fifteen new PV-2 *Harpoons*.

Between 23 and 30 June, squadron crews ferried the PV-2s to the Lockheed plant at Burbank, California, for wing spar modifications. This delayed the squadron's departure for a third tour of duty in the Aleutians. With the return of the first three planes from Burbank, orders were received to proceed to Attu. Beginning 4 August 1945, the squadron departed in increments of three planes, the last plane arriving on Attu 22 August 1945, six days after the end of WW II. VPB-135 and their *Harpoons* had been scheduled to furnish air support for a planned invasion of the Kuriles by Russian troops, who were in training at Cold Bay at the time. An event which, in historical reflection, thankfully did not take place.

Lt. J. P. Vivian's crew preparing for a mission.

U.S. Navy

VB-135 crew donning heavy flight clothing and survival gear. Crews were issued Smith & Wesson .38 cal. revolvers for side arms. "Butch" Mason's 3V3 Ventura in background.

Lockheed

PV-2 over Los Angeles area, May '45. Number on fin is a company designator and Harpoon on the nose applied for publicity. Plane was later delivered to VPB-135 at NAS Whidbey Island.

SOME VB-135 CREWS

USN via M. A. Mason

**The crew of Lt. M. A. "Butch" Mason, Executive Officer of VB-135.
Mason led the first daylight mission on 12 June 1944.**

**Back row, l. to r.: Mason, Bernhardt, Hanlon
Front row: Aho, Horstman, McGee, Tiner**

USN via M. A. Mason

**The most decorated Empire Express crew.
L. A. "Pat" Patteson (two DFCs and two Air Medals)
The entire crew received Air Medals and Letters of Commendation for completing two daring solo photographic runs over Shimushu and Paramushiro on 14 and 25 June 1944.**

**Back row, l. to r.: Patteson, Walsh, Rice
Front row: Hillard (AMM) Cach (ARM) Jacobsen (AOM), Britt (AMM)**

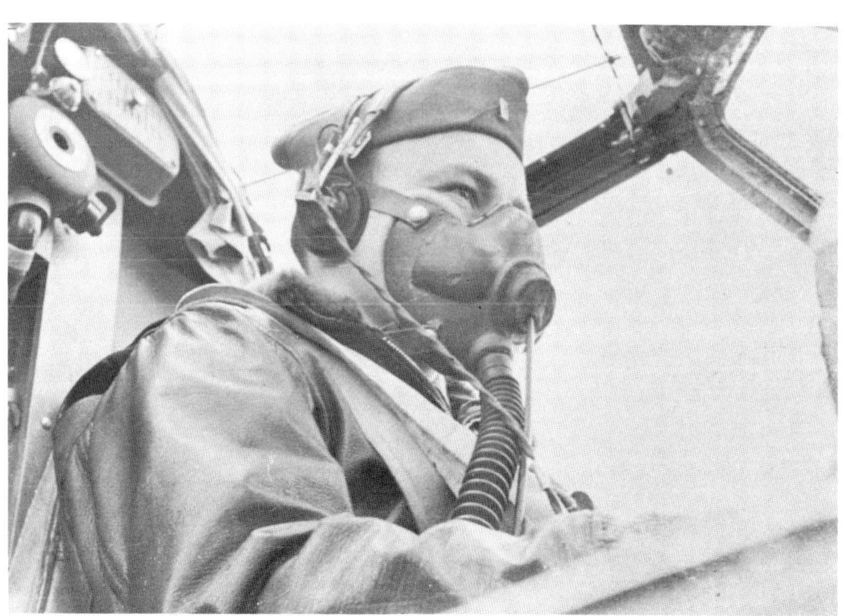

L. A. Patteson

Lt. L. A. "Pat" Patteson at the controls of his PV-1.

Lt. J. P. Vivian's crew, VB-135. Their PV received battle damage after attacking and sinking an armed picket boat off Shimushu on 23 July '44 and they were forced to land at Petropavlosk, Russia. Their plane was impounded and the crew made virtual prisoners. They were eventually returned to the U.S. via Europe. Back row, l. to r.: Wilson, Vivian, Edwards. Front row: Virant, Nommensen, Anderson, Schasney.

U.S. Navy

VB-135 crew: Back row, l. to r.: Brownlee, Lester, Rice. Front row: Landsdowne, Lamb, Denison, Segar.

USN via M. A. Mason

VB-135 crew of Lt. J. W. Pool who, on 23 July '44, were attacked by eight enemy Tojo type fighters. Lt. Pool shot down one with the bow guns. Their PV received heavy battle damage but landed at Attu successfully by using the emergency hydraulic system. Back row, l. to r.: Pool, Reilley, Warner. Front row: Kreisle, Haycraft (DFC), Scott, Marker, Lee.

U.S. Navy

VB-135 crew, Back row, l. to r.: Mabus, Spiva, Schuster. Front row: Bowman, Woodworth, Wood, Cassidy.

USN via R. R. Larson

Lt. Blakeney and crew after making the first daylight flight over the Kuriles.

U.S. Navy

VB-135 crew, Back row, l. to r.: Lt. J. F. Rumford, Gardner, Robillard. Front row: Bradbury, Hewitt, Santoro, Arkins.

History of Bombing Squadron - 136

VB-136

Bombing Squadron 136 was commissioned 1 March 1943 at Ault Field, Naval Air Station, Whidbey Island, Washington, under the command of LCdr. Nathan S. Haines. The training period for the squadron at Whidbey Island was from 1 March to 23 April 1943. The squadron was composed of eighteen five man crews. All the patrol plane commanders had previous operational experience in PBYs, but only an average of 25 hours in the PV-1 before assignment to VB-136. Training included: navigation flights, a limited amount of instrument flying, gunnery, torpedo launching, radar, glide and low-level bombing. All the PV-1s delivered to VB-136 were, as were the original British production, of single-pilot cockpit configuration. All flight instruments were on the left side of the cockpit — no co-pilot controls or seat were fitted. only a folding canvas affair. Dual controls were furnished as a kit installation in the field. Radar was also installed in the aircraft at Whidbey Island. The fuel system was a nightmare, 13 separate fuel cells, all manually fed. While the pilots were experienced they constantly wrestled with keeping their nerve and composure. Their transition to the PV-1 was justifiably difficult.

On 23 April 1943, the squadron departed Whidbey Island for Adak, arriving there on 30 April 1943. The squadron immediately started operations including searches and missions against Japanese held Kiska. Extensive anti-submarine missions were performed in support of the subsequent invasion of Kiska. During the month of September routine search operations proved negative.

Patrol duty was hours and hours of boredom, punctuated only now and then by moments of stark terror! On one such occasion Lt. (jg) King was flying at a mere 35 feet altitude, making an instrument approach to Adak when suddenly he discovered an object on his radar. There should not have been any vessels in that vicinity at the time. He quickly executed a sharp pull-up. In doing so he caught one wing in the water. The plane gave a tremendous lurch and begin shaking badly. However, it kept on flying and remained in complete control. After making another approach and landing, inspection revealed that the entire empennage was severely damaged, all the rivets had been popped or sheared. The fairing alone was holding the skin in place.

On 1 October 1943, the entire squadron was moved to Attu for subsequent operations, including searches and offensive air sweeps against enemy aircraft from the Kuriles. On the offensive sweeps, contact was made on two occasions with Japanese *Bettys*. Lt. (jg) Connors picked up an aircraft blip on the radar one day near the end of his outbound leg of a sector search. When the blip became visible it proved to be a *Betty*. The PV gave chase, gaining on the enemy aircraft as it headed for home at full speed. Connors gained steadily but the two thirsty R-2800s were gulping gas at an alarming rate. With an eye on the fuel guages they reluctantly had to abandon the chase after 15 minutes. In a similar encounter Lt. Dinsmore had a twenty minute running battle with a *Betty* and managed to damage it. He burned out his bow guns in the process though.

Squadron VB-136, Naval Air Station, Whidbey Island, May 19
Group departed for Attu shortly thereafter.

VB-136, PV-1 enroute to Paramushiro, 1944

USN via R. R. Larson

On 16 November 1943, Lt. H. K. Mantius and crew was the first FAW-4 plane to conduct an offensive sweep to Paramushiro and the *Empire Express* was born. On 10 December 1943, VB-136 was relieved of duty by VB-139 and departed for Seattle.

VB-136 was subsequently reformed at NAS Whidbey Island under the command of LCdr. Charles Wayne. Their PV-1s were greatly modified with new equipment installed, such as LORAN, a dual instrument panel, a Bendix ADF receiver and VHF radio, an ARC receiver with anti-static loop, a radio altimeter, an aft cabin fuel tank and an automatic fuel system. The oxygen system was improved and the chin armament consisted of three .50 cal. guns. The training period extended from 1 March to 6 June 1944. On 7 June the squadron departed for Attu, arriving on 17 June 1944, for subsequent operational searches, tactical bombing and photographic missions over the Kuriles.

The target area was 650 to 700 miles from Attu and except for neutral territory in Russia, there was no alternate field between. Even minor damage offered but two choices—making an emergency landing in Russia where the plane would be impounded and the crew imprisoned or limping back over 700 miles of freezing water. The great distance to the target obviated the feasibility of fighter escort. This serious problem was demonstrated when Lt. Price, while damaging a 5000 ton freighter off Paramushiro, was hit by flak in the port engine, aileron and wing. Oil pressure in the engine dropped, all unnecessary gear was jettisoned and single-rescue PBYs were sent out and although one was eventually behind them no actual contact was made. The *Ventura* was able to return only because the damaged engine operated until the last hour of flight. A tribute to the Pratt & Whitney engine.

It was decided that a PV-1 escorting another crippled PV-1 home could be helpful especially in trying weather conditions. A PBY could help if a ditching was necessary. In a follow up order a rescue PV and crew was subsequently kept on a 15 minute alert status. Two medical kits and a 7-man life raft, equipped with parachutes, were carried in the bomb bay. Fortunately the droppable rescue flotation gear was not used but the alert-plane was called upon several times to assist other stricken aircraft in navigating home.

By 1944, Attu was a greatly improved base, living conditions were far better than during VB-136's first deployment there. Officers lived six to a Quonset hut with privately partitioned rooms. The enlisted men lived twelve to a Quonset. Much of the furniture was built by the inhabitants but some "stateside" furniture made its way to Navy procurement. Morale was better than ever before and interest in the various sports programs was good. Basketball, handball, ping-pong and even chess tournaments were underway.

The effectiveness of 136's punch was impaired by the type of bomb load they normally carried. Improvements were undertaken. Headquarters FAW-4 issued an order . . . rather than load three 500 pound general purpose (GP) bombs, a mixed load of one 500 pound GP, one fragmentation cluster

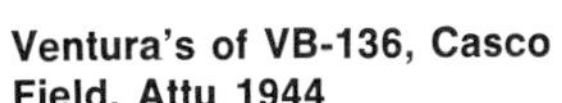

Ventura's of VB-136, Casco Field, Attu 1944

USN via R. R. Larson

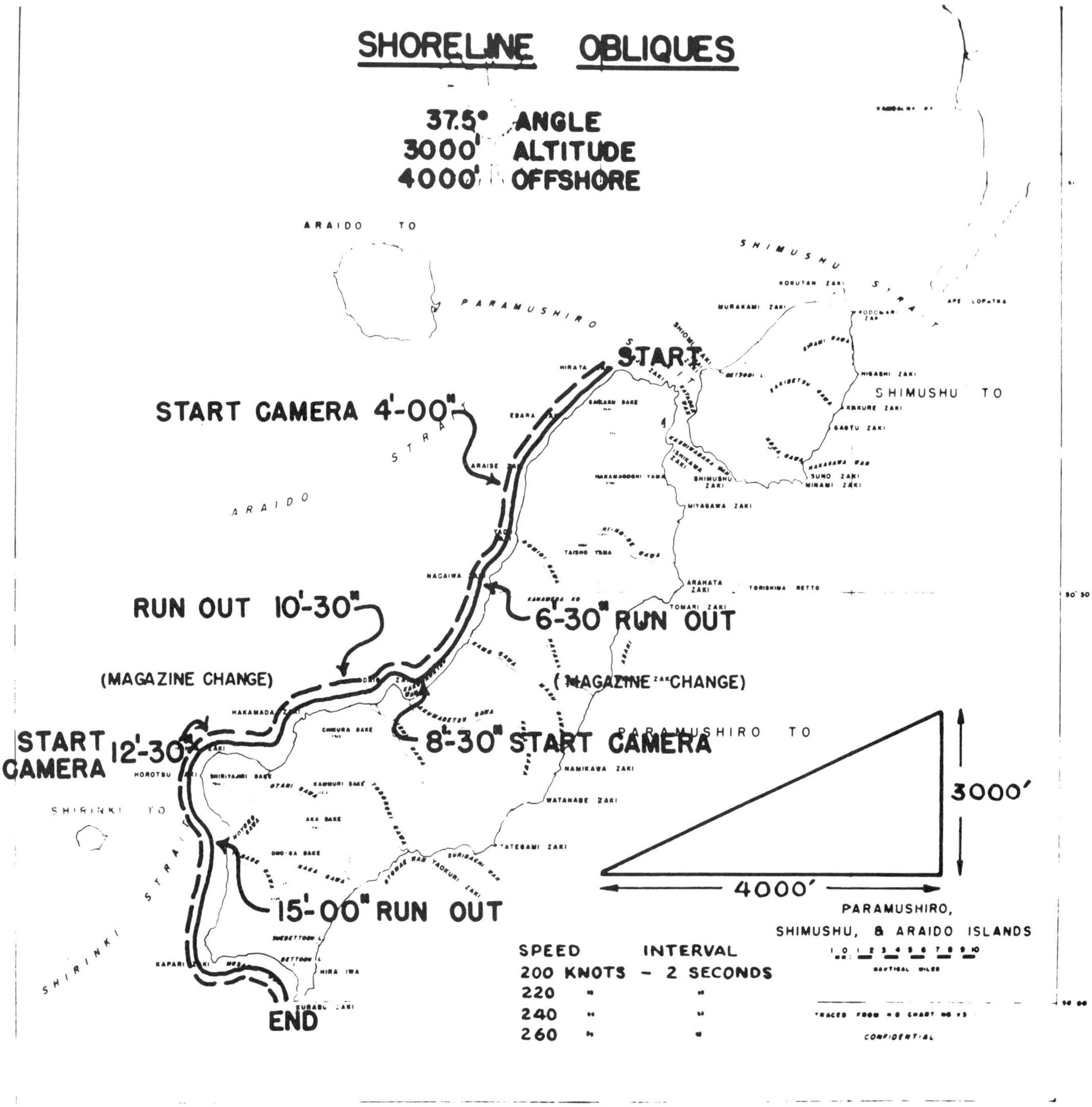

SHORELINE OBLIQUES

37.5° ANGLE
3000' ALTITUDE
4000' OFFSHORE

ARAIDO TO
SHIMUSHU STRAIT
PARAMUSHIRO STRAIT
ARAIDO STRAIT
START
START CAMERA 4'-00"
SHIMUSHU TO
RUN OUT 10'-30"
(MAGAZINE CHANGE)
6'-30" RUN OUT
(MAGAZINE CHANGE)
START CAMERA 12'-30"
8'-30" START CAMERA
PARAMUSHIRO TO
SHIRINKI TO
SHIRINKI STRAIT
15'-00" RUN OUT
END

3000'
4000'

PARAMUSHIRO, SHIMUSHU, & ARAIDO ISLANDS

SPEED	INTERVAL
200 KNOTS	2 SECONDS
220 "	"
240 "	"
260 "	"

CONFIDENTIAL

The PV-1 of Lt. Price, VB-136, after a single engine landing at Attu. Port engine was hit by AA fire off Paramushiro. The disabled engine continued to run until the last hour of flight. Oil pressure finally became so low that Price had to cut the engine and feather the prop.

USN via R. R. Larson

and one incendiary cluster, all with delayed action fuses, were to be regular offensive armament. In theory the were to look for a target for each type of bomb and score a direct hit with each. In making attacks on shipping, the bombs were to be dropped in a sequence using the intervalometer but only through luck would the 500 pound GP score a hit.

On 12 September 1944, Lts. Moorehead and Bacak made visual landfall on Kamchatka from 120 miles out. Dropping to the deck both continued westward until off the east coast of Shimushu. Lt. Moorehead made his bombing run on the Sakibetsu fishery building from 50 feet at 240 knots. His incendiary load failed to release but his GP bomb was seen to strike among the buildings. During the run he strafed several buildings and a gun position west of the fishery. Meager and inaccurate AA fire was encountered from the area and a few bursts of heavy AA were received from batteries at Miyoshino as he circled inland.

Suddenly seven radial engine *Hamp* fighters were sighted. The first two approached from astern about a mile away. They followed for ten minutes but were eventually outdistanced when Lt. Moorehead rammed the PV to 280 knots (322 mph). Shortly after the first sighting three *Oscars* were also sighted at about 1000 feet altitude and one mile to starboard. The PV was streaking at 50 feet altitude at the time when one of the fighters began a run from 4 o'clock. As he neared the bomber he had second thoughts, broke off and turned away without firing. All three fighters then followed suit. Above five minutes later two more fighters were sighted flying at the same altitude and about a mile distant. Lt. Moorehead increased speed to 280 knots again, the fighters followed for 20 minutes but by this time they were unable to overtake the PV. As they fell behind, the fighters opened fire and several splashes were seen in the water behind the PV.

On 14 September the planes of Lts. Morrison and Littleton attacked targets on the eastern coast of Paramushiro. Lt. Morrison attacking a cannery at Hayake Gawa and Lt. Littleton, a fishery at Kasuga Zaki. Lt. Morrison flew northeasterly at 50 feet altitude along the coast of Paramushiro. Crossing the coastline he circled inland and climbed to 1,000 feet. From here he made his run to seaward in a low-level attack, greasing the deck at 100 feet and at speeds up to 280 knots. After completing the run he was jumped by two fighters, who gave chase for five minutes. They were unable to close however.

Lt. Littleton dropped his bombs from 200 feet on the fishery building at Kasuga. The bombs were seen to strike the buildings but they were immediately attacked by four single engine fighters which were first sighted to starboard about one half mile distance. He turned to starboard thus bringing two of the fighters in on him from dead ahead and above. He fired at the lead plane with the bow guns causing the fighter to pull up and expose its belly. A second burst caught it and it exploded in flames, passing over the PV. It was picked up and hit again by the turret gunner and fell into the sea. The second fighter also came straight in from 12 o'clock, streaking in a blur over the PV at about 50 feet and firing all the way. Littleton took advantage of the PV's greatest attribute, Firewalling the throttles, he dived to 20 feet altitude and out outdistanced the remaining fighters. The crew thought the enemy planes resembled *Oscars*, except that at least four guns were seen to fire from the wings of the two planes which made direct attacks. After evaluation of the photographs of the downed fighter intelligence listed the plane as a *Tojo*.

U.S. Navy

TOJO shot down by Lt. F. R. Littleton, 17 Sept. '44.

U.S. Navy

VB-136 PV-1 landed high and long. Unable to stop on the icy runway it slid off the end.

Kakumabetsu airfield photo taken by Lt. C. B. Nelson, VB-136, during a bombing run.

U.S. Navy

enemy. One instance when the "bubble" collapsed with planes in the air was on 2 July 1944. A return to base order was sent to all planes because of detiorating weather. Lt. Hayes arrived over Attu and found the fog solid from 1500 feet down to the water. He was one of six planes held in the clear, prior to instrument let-down clearance. During the hour that Lt. Hayes and Lt. Larson, in another PV, held over the high cone, one plane landed at Alexai Point and two at Casco. A fourth PV squeezed into Shemya. Lt. Hayes tried his first approach at 150 feet on radar and had to pull out. A second approach at 50 feet also failed. As an old timer on Aleutian let-downs Lt. Larson should have made it but his radar operator was inexperienced and could not recognize the land forms on his scope. Larson passed the crossing leg to the left of the low cone and started timing his final approach across Massacre Bay. At that crucial moment the range was turned off! This really rattled him and he started a climbing turn to the left to stay away from the mountains. He was advised by the tower that the weather was deteriorating too rapidly for further attempts.

Lt. Hayes tried Shemya three times at 150 feet, 100 feet and 50 feet with no success. As he started his last approach he heard from Larson, who was going to try Amchitka. He followed suit jettisoning all unnecessary gear enroute. Amchitka was reporting ceiling and visibility improving .Lt Larson described his landing at Amchitka, "We made the high cone and started to make the let-down and procedure turn. We had a radar altimeter that read true altitude above the terrain and that instrument was really valuable that night. I started timing the final leg after identifying the low cone. I let down to 100 feet on the radar altimeter, but when my elapsed time was up I couldn't see a thing. One set of lights flashed by which I later realized was the control tower. We aborted the let-down attempt and climbed on top again. The control tower operator then asked if I wanted the field landing lights on! It was small wonder I couldn't see anything. I told him to get those lights on fast. It turned out they were brand new experimental high intensity Bartow lights that were being evaluated in low visibility conditions. I saw those beautiful lights half a mile away on the next pass and would have cheerfully paid for every one of them with my own money! We landed, finally."

Lt. Hayes missed his first attempt at 100 feet. His second approach was made at 50 feet over the water and continued at 50 feet over land by radar altimeter. Barely missing a hangar they spotted the runway as they started around and ploughed onto the runway at 120 knots with no flaps, landing safely.

On 17 September the PV of LCdr. Wayne, the squadron skipper, was damaged over Paramushiro and made a forced landing in Russia. After this tragic loss, Commodore Gehres felt that the squadron had suffered enough losses and cancelled further *Empire Express* missions for VB-136. After that date 136 was assigned sector searches or special photo missions where the speed of the PV-1 was required.

In the fall of 1944, invasion of the Kuriles by allied ground forces was in the definite stages of planning. With their long experience and vast knowledge of the area, the Navy was called upon to gather much needed intelligence on suitable landing beaches. FAW-4 headquarters received orders to photograph the western shores of Paramushiro and Shimushu. To successfully carry out this mission in a minimum amount of time the Navy devised an all out effort. The plan consisted of ten PV-1s which would create a diversion by attacking Karabu Zaki and Shimushu, while five other PV-s equipped

U.S. Navy

Japanese freighter under attack by the bow guns of Lt. F. R. Littleton's PV-1.

Lockheed

The Navy brought the PV-1 Ventura to the Alaskan area because of its speed and increased firepower.

with large F-56 cameras would cover the Paramushiro coastline, obtaining high and low oblique photographs. With fifteen PVs up at one time on a high risk mission, two PBY-5A "Dumbo" aircraft of VP-62 were to stand by off the east coast of Kamchatka for rescue operations.

The mission was attempted on 2 November but was forced to return due to low cloud cover over the areas to be photographed. Finally on 6 November Lts. Morrison, Jones, Bacak and Moorehead of VPB-136 succeeded in obtaining the photographs. Four PV-1s of VPB-131 made the diversionary attacks on the east side of the island. The mission did not go as originally planned but the final results were the same. The areas was heavily defended by enemy fighters and anti-aircraft guns but all eight planes returned safely.

Late in November VP-43 departed Attu and VPB-136 took over. 136 engaged in sector searches during the remainder of their stay at Attu while VPB-131 assumed the intruder role.

By 12 March 1945, all fifteen PV-2 *Harpoons* of VPB-139 were on Attu and VPB-136 was officially relieved as of that date. During their second deployment in 1944 VB-136 lost three planes and crews to enemy action and one to mechanical failure. All four made forced landings in Russia and the crews were eventually returned to the States.

Lt. Moorehead was the only one to have further trouble. He ran into problems while ferrying his plane back to the U.S. While attempting to home-in on the Kodiak radio range, he became lost. The low-frequency ranges at that time had the bad characteristic of "swinging" near sundown. He was forced to ditch in the water off Karluk. The crew managed to get the life raft out and inflated and all hands were saved even though a passenger aboard could not swim. One good thing came of the ditching. It was generally held that Al Moorehead had a set of "Aviation Greens" uniform . . . that was the worst color in the U.S. Navy. It went down with the airplane and Al was forced to buy a new uniform. He was then accepted back by shipmates. (During WW II the Navy had *optional* Aviation Green uniforms for Aviation officers and Chiefs.)

In May 1945 VPB-136 was reformed with PV-2 *Harpoons* but the end of the war made it unnecessary to return to the Aleutians.

Lt. Larson's turrett gunner, Aircrewman Brooks, cleans the plexiglass blister using mild soap, water and soft rag. taking care not to scratch the delicate surface.

USN via R. R. Larson

SOME VB-136 CREWS

VB-136 crew of Lt. F. R. Littleton. Lt. Littleton was credited with shooting down a Tojo fighter. Back row, l. to r.: Scott, Littleton, Anderson. Front row: Peterson, Zwingli, Hadley.

Lt. Morrison's VB-136 crew. Back row, l. to r.: Davison, Morrison, Waggoner. Front row: McLane, Clement, Roehren.

VB-136 crew of Lt. Robert Larson. Back row, l. to r.: Jernigan, Larson, Loring. Front row: Farmer, Brooks, Kelly.

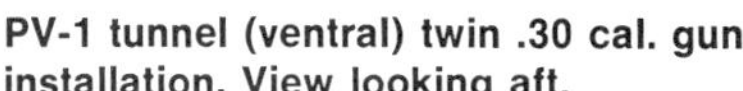

PV-1 tunnel (ventral) twin .30 cal. gun installation. View looking aft.

A. A. Hoffman, AOCS

RE-OCCUPATION OF KISKA

Allied invasion force hit the rugged terrain of Kiska.

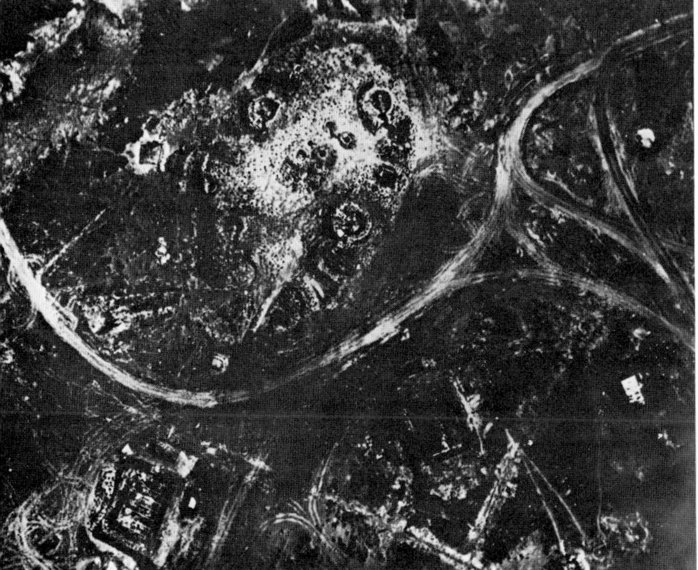

Four days after reoccupation, 8-26-43, shell potted tundra clearly show abandoned Japanese gun installations.

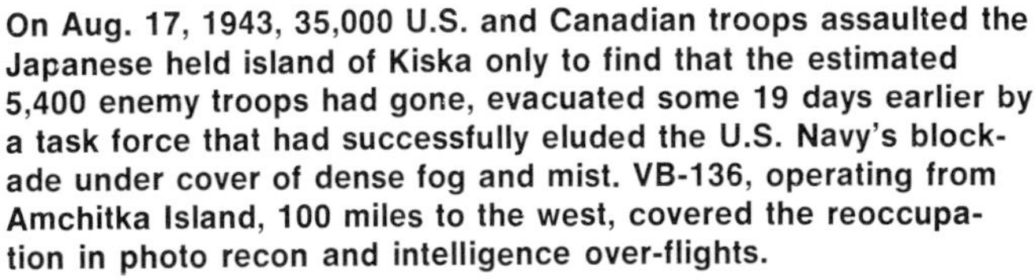

On Aug. 17, 1943, 35,000 U.S. and Canadian troops assaulted the Japanese held island of Kiska only to find that the estimated 5,400 enemy troops had gone, evacuated some 19 days earlier by a task force that had successfully eluded the U.S. Navy's blockade under cover of dense fog and mist. VB-136, operating from Amchitka Island, 100 miles to the west, covered the reoccupation in photo recon and intelligence over-flights.

All Photos U.S.N. (VB-136)

History of Bombing Squadron - 139

VB-139

Bombing Squadron 139 was commissioned 1 April 1943, under the command of LCdr. W. R. Stevens. Commissioning ceremonies took place at Ault Field, Naval Air Station, Whidbey Island, Washington, where training began. On 22 July 1943, the squadron was transferred to Naval Air Station, Alameda, California, for the new instrument panel installation in their PV-1s. The squadron returned to Whidbey Island on 15 August 1943, for further instrument training.

On 1 October 1943, the squadron departed Whidbey Island for the Aleutians in three, five-plane divisions. Bad weather and mechanical difficulties were encountered enroute. Twelve planes finally arrived at Amchitka on 7 October 1943, and patrol missions were undertaken the very next day. The remainder of the squadron finally arrived by 10 October. During this time the squadron operated basically as a Patrol unit. Missions were characteristically uneventful. The weather constituted a serious hazard, delaying operations at times and often seriously impairing completion of missions.

The entire squadron, with the exception of three crews stationed at Adak, moved to Attu on 10 December. They relieved VB-136 in routine patrol and search duties until 19 January 1944. Sector searches rotated within the squadron and extended from 400 to 550 miles. Usually each crew flew one day and had the following two days off.

Although the squadron, after the 19th of January, still flew occasional patrols, its primary function became night reconaissance and the bombing of Paramushiro and Shimushu. An interesting side note of this period is that it was during this time that VP-43 began their legendary night photo-recon flights over the Kuriles flying PBY-5A *Catalinas* in the "Black Cat" fashion.

It was Lt. Mantius of VB-136 who demonstrated the PV-1s capability of making flights as far as the Kuriles. On 19 January 1944, Lieutenants R. A. MacGregor, D. M. Birdsall and T. H. McKelvey made the first night flights over the same area in PV-1s. Together they formed the nucleus of a tactical offensive, demonstrating the feasibility of the PV-1 as a long-range medium bomber to strike the northern Japanese homeland.

For the next four months VB-139 was primarily engaged in night *Empire Express* runs. The following is a typical night mission taken from the squadron war diary:

"Takeoff at 1156 hours. Clouds were topped at 8500 feet, the pilot going to cruising settings of 1800 rpm and 32" Hg. At 1520 obtained radar landfall on Cape Lopatka: at 1533 passed over the Cape and crossed Shimushu. The area had heavy cloud cover so the pilot took a westerly heading over the Sea of Okhostsk, then turned southwest until a radar fix was obtained on Shimushu. The course was changed to 190° for a run over Kakumabetsu. The target area was 8/10 covered at 10,000′ but the photoflash bombs were dropped through a hole and obtained a clear picture of Kakumabetsu. No enemy activity was observed, the pilot continuing toward Karabu Zaki. On a north-south pass over Karabu all remaining bombs were dropped through the overcast. They departed Karabu at 1605 hours with 720 gallons of fuel aboard. It had been necessary to fly through frontal conditions going to the target. On the way back the air speed indicator froze. Moderate to severe icing conditions were encountered until an hour from base. Landed at 2024 hours with 260 gallons of fuel aboard".

Heavy anti-aircraft fire was frequently encountered although very inaccurate. One plane and crew was lost on one of the night missions. VB-139 successfully executed 78 out of 100 attempted photographic and bombing missions over the Kuriles. These were made under sub-arctic winter conditions with its attendant hazards of ice-covered runways, storms and targets 700 miles across the freezing North Pacific. Throughout these operations the squadron functioned independent of any other unit. Its sole contact with Army squadrons based on Attu consisted of assistance in the training of their navigators. Other than the one crew which was lost during a mission, four crewmen were killed in a takeoff crash at Attu. Three other planes were lost in operational crashes but all the crews survived. Planes were parked in revetments and despite the extereme adverse conditions under which they worked, maintenance crews managed to have the planes ready for all operational missions.

Having served six tough winter months in the frozen north, VB-139 was relieved of duty on 5 May 1944, and departed for the United States to assume further assignments. Their duties were taken over by VB-135.

VB-139 was recommissioned on 1 August 1944, under the command of LCdr. Glenn A. David. Training was started at

U.S.N.

Venturas of VB-139 in Attu revetment — early 1944.

Whidbey Island with twenty per cent of the former officers and enlisted men being retained, thus furnishing a background of experience for the new members. Primary emphasis was placed on G.C.A. (Ground Controlled Approach) and rocket attacks. Casco Field on Attu had been equipped with this new form of landing control. Proficiency in its use was mandatory under the low-ceiling conditions encountered in the Aleutians.

Early in February 1945 the squadron began receiving their fifteen PV-2 *Harpoons*. Familiarization flights in the new ships were accelerated, but serious problems began to manifest themselves in the planes and HEDRON worked day and night to correct the bugs.

On 1 March 1945, a six-plane section departed for Attu. Two days later the remaining nine *Harpoons* departed. On 12 March 1945, VPB-139 officially relieved VPB-136's duties in the Aleutians. The squadron began flying sector searches two days later. These routine missions continued until 28 March. When a four-plane section attacked Kokutan Zaki with rockets on 6 April the PV-2 *Harpoon* officially joined the *Empire Express*.

10 April: a four plane section attacked the cannery at Hayake Gawa with rockets. Many hits were observed and all three buildings were left burning.

15 April: Four *Harpoons* attacked Tomari Zaki cannery, scoring eight hits on the buildings. One of the planes attacked a picket boat with machine gun fire, expending 350 rounds. The boat attack was disrupted by three fighters but they did not press their attack.

28 April: four *Harpoons* attacked Minami Zaki radar installations. Eleven rocket hits were observed within the target area, and approximately 2000 rounds of .50 caliber fire were expended. Two of the planes were hit by AA fire but damage was minor.

After April 1945 the wing spars of the *Harpoons* were deemed inadequate. A Bureau of Aeronautics order restricted their operational use to patrol, reconnaissance under weather conditions permitting a ready escape from enemy fighters and rocket attacks on shipping.

10 May: eight *Harpoons* attacked radar installations at Minami Zaki in two sections of four planes each and at two hour intervals. Both sections were greeted by AA fire, necessitating a level approach to the target. Close observation of rocket hits was prevented due to the intense ground fire. The first section set buildings on fire and enemy personnel were attempting to extinguish the fires when the second section arrived. Five of the eight PV-2s were hit by AA fire causing major damage to three. Two crew members were injured, neither requiring hospitalization. Two others had bullet holes in their sleeves and trousers.

As the weather improved, missions increased with a corresponding increase in enemy fighter defense activity. Fighter opposition was encountered but it was not very aggresive.

12 May: two *Harpoons* made an anti-shipping sweep and reconnaissance of the west coast of Shimushu and Paramushiro. The entire target area was closed in by a solid overcast. Shimushu was crossed twice and the Okhotsk Sea was searched by radar. All rockets were expended at Minami Zaki and aimed by radar.

14 May: two *Harpoons* searched the west coast of Shimushu and a large area of the Sea of Okhotsk. Results were negative until they were about seven miles northwest of Kataoka harbor, where two armed trawlers were sighted and attacked. One rocket was seen striking one of the ships.

18 May: reconnaissance of Paramushiro Straits was conducted from the west coast of Shimushu by two *Harpoons*. Both planes sought cloud cover in the overcast when fighters were sighted, then dropped below for their rocket and bomb

When Lt. R. E. Garnett developed an oil leak over Shimushu, he risked flying 700 miles back over the freezing ocean on one engine and landed his PV-2 perfectly on Attu, with oil stained nacelle and feathered prop.

runs. Each plane carried eight 5″ HVAR rockets and three 250 lb. bombs. One *Harpoon* was hit in the port chin-gun by a 20mm projectile. The fighters were lost after leaving the target area.

19 May: two *Harpoons* on an anti-shipping sweep searched the Sea of Okhotsk with negative results. Shortly after turning for the shore line search a single enemy fighter was sighted paralleling their course. The fighter was able to close and made a run on the trailing *Harpoon* but the attack was broken off shortly thereafter.

20 May: two *Harpoons* searched the Sea of Okhotsk and Paramushiro Strait. Departing from the area both planes fired their rockets at Kokutan Zaki installations.

26 May: four *Harpoons* searched the east and west coasts of Shimushu and Paramushiro, splitting into two sections at the target. The first section proceeded to survey the grounded Russian freighter *Mariupol*. Some sort of unusual activity had been reported on or around the stricken ship. Upon approaching the area, one aircraft observed a trawler and attacked with the bow guns. The *Harpoons* rendezvoused when four fighters were sighted. They did not close and finally disappeared from sight.

3 June: two *Harpoons* searched the west coast of Paramushiro. Bombs weer dropped through the overcast on Kataoka.

During the month of June more operational missions were flown than in any previous month, consequently maintenance became more of a problem because the PV-2s were still being flown under certain restrictions pending structural reinforcements and major overhauls.

4 June: two *Harpoons* took off on an anti-shipping sweep. Bombs were dropped on Hayaka Gawa by radar. Hits were believed to be scored.

8 June: two-plane sections of *Harpoons* went on an anti-shipping sweep and reconnaissance of the east and west coasts of Paramushiro. Flight Able, led by Lt. MacGregor, searched the east coast and dropped bombs on Katoaka through the overcast. Flight Baker, led by LCdr. David, received AA damage from an unidentified source between Cape Lopatka and Shimushu.

9 June: four *Harpoons* went on an anti-shipping sweep. Four trawlers south of Masugawa were bombed and strafed. During the attack, moderate AA fire was received. All planes returned, two had received minor damage.

10 June: two two-plane sections of *Harpoons* completed an anti-shipping sweep of the east coast of Paramushiro. Offshore, Lt. Torry and Lt. (jg) Bradbury observed an enemy freighter. Soon they were joined by Lt. Bampus and Lt. (jg) Heinken, at which time strafing and bombing runs commenced. The freighter was left burning and out of control.

17 June: four *Harpoons* went on an anti-shipping sweep. Flight Able searched the Okhotsk Sea and Flight Baker searched the east coast of Paramushiro. All planes bombed Suribachi by radar, the only target of value they could find.

18 June: four *Harpoons* took off on an anti-shipping sweep. Flight Able searched the Okhotsk Sea. Flight Baker searched the east coast and bombed Tomari Zaki.

22 June: four *Harpoons* searched the east coast of Shimushu and the northern portion of Paramushiro. The search was interrupted by three *Oscars* which were sighted dead ahead, coming out of the sun. Tight 180° turns were executed and after a period of five minutes the *Harpoons* were able to take refuge in scattered cloud cover. All planes returned safely.

24 July: a detachment of six planes and nine crews departed for Amchitka, where they flew sector searches.

8 August: seven planes of VPB-135 landed at Attu.

10 August: five more *Harpoons* of VPB-135 arrived, and VPB-139 was relieved of duty.

20 August: the remaining twelve *Harpoons* of VPB-139 departed for Whidbey Island.

The squadron had lost three planes and one crew during their second tour of duty. On 27 March one plane crashed on landing at Shemya, was totally destroyed by fire, but the crew escaped. One plane and crew failed to return from a search mission on 22 April and was never heard from again. One plane was extensively damaged by AA on 10 May but was able to return to Attu.

By 23 August 1945, the last *Harpoon* landed at Whidbey Island and on 13 September Patrol Bombing Squadron 139 was officially decommissioned.

OME VB-139 CREWS

USN via L. A. Patteson

Loading a torpedo aboard a PV-2 during a bitter snow storm. Hedron crews worked under the most miserable conditions imaginable.

USN via F. N. Murcray

Crew of Lt. R. A. McGregor, one of the true pioneers of the Empire Express. He was Operations Officer of VB-139 during their first deployment with the PV-1 Venturas and the second deployment in PV-2 Harpoons. Lts. McGregor, T. R. McKelvey and D. M. Birdsall formed the nucleus of a tactical offensive through a thorough understanding of the PVs capabilities and limitations. They proved that the plane could be used to strike distant Japanese installations in the Kuriles. Lt. McGregor is left in back row.

USN via F. N. Murcray

Crew of LCdr. William R. Stevens, skipper of VB-139 during their first Aleutian tour of duty. LCdr. Stevens, back row left. Unusual crew composition, an officer and enlisted navigator and the ARM is a Chief.

History of Patrol Bombing Squadron - 131

VPB-131

Patrol Bombing Squadron 131 was the final PV-1 squadron to join the *Empire Express*. It was commissioned at the Naval Air Station, DeLand, Florida on March 1943 where twenty-four officers and fifty-one enlisted men were assembled in the main hangar in dress blues. The orders placing the squadron in commission were read by LCdr. B. T. Talbett, placing LCdr. John A. Gamon, Jr. in command. Pictures were taken of the squadron and at 1215 that day flight operations commenced. All the pilots were from PBY *Catalina* and OS2U *Kingfisher* inshore patrol squadrons. None were familiar with the PV-1 *Ventura* and for the next three weeks the crews attended lectures on the PV, while awaiting arrival of the aircraft. Finally on 15 March the squadron received two PV-1s. After four of the pilots had been checked-out they began checking-out others. From then until the end of May the squadron trained in bounce-bombing with waterfilled bombs, advanced navigation and night flying. Since the squadron was destined for anti-submarine work, this part of the training was stressed.

Although the squadron had received only four of its aircraft, it departed on 31 May for NAAF, Boca Chica, Florida. Here the squadron trained in strafing, bombing and night navigation with the "S" type submarines based at Key West, making practice runs on them during the daylight hours. During this training period, one of the aircraft was reported overdue from a routine bombing hop west of Key West. Although search operations were conducted for the next three days, no trace was ever found of the PV-1 or its crew.

The squadron finally received its twelve *Venturas*, some new, others not so new, and on 28 June 1943, departed for Naval Air Station, Guantanamo Bay, Cuba, as an operational squadron under Fleet Air Wing Eleven. They operated as a patrol unit from Guantanamo Bay, San Juan, P.R., Trinidad, B.W.I., and Dutch Guiana while in the Atlantic, returning to Norfolk, Virginia in March 1944.

Two months later VPB-131 was reformed at Whidbey Island under the command of LCdr. Rolland Hastreiter, USNR. The Army Air Force resumed bombing the Kuriles late in 1944 and the Navy wasn't to be left out. The Navy would enter the battle of the Aleutians with a revolutionary new weapon — Aircraft Rockets! VPB-131 would carry the Navy banner and to display their ability, rocket supports and the five gun "Chin Gun" package was immediately installed on the PV-1s. LORAN and the more modern ASD-1 radar was also put aboard. Deck armor aft of the center of gravity was removed and bomb bay tanks were installed for increased range. Since 131 was to undertake the intruder role, the tunnel guns were retained for strafing purposes. Under these modfications weight distribution remained critical. On takeoff all tunnel and turret ammunition was brought forward to the radio compartment and the entire crew crowded into the same area, where they remained until airborne and a stable flight attitude attained.

Coming from the sunny climate of the Caribbean, VPB-131 lacked experience in cold weather operations. A great deal of discomfort was experienced and they had some difficulty in getting acclimated. The squadron began an intensive training program concentrating on radar, glide and masthead bombing. Despite all this they were destined never to carry a single bomb. Primary emphasis changed to rocket-firing and strafing. During their training they had to use old British made rockets. This was unfortunate because their performance left much to be desired and the crews were left with a certain amount of apprehension about what the outcome would be once they were in actual combat. More advanced and reliable American versions were issued later.

LCdr. Hastreiter was an exceptional skipper and a great aviator. He was aggressive but cautious and earned the respect of every man in the squadron. Before they left Whidbey Island he flew in the cockpit with every crew. In this way they learned from each other but the Commander's purpose was to acquire a solid feel for each crew's individual capabilities. Later during the pressures of combat, he always had a good grasp of what he could expect from his men. Each crew made its first mission over the Kuriles personally led by LCdr. Hastraiter. Never did they go alone nor with another leader until they met with his satisfaction. His PV-1, number 85V, was referred to by his chief petty officers as "A Flying-fool's Machine", and bore the dings and scars of battle.

The squadron departed Whidbey Island on 8 October 1944. After refueling stops at Annette, Yakutat, Kodiak and Umnak they arrived at Attu on the 20th. At Umnak,

Lockheed PV-1 in "Atlantic" colors, Gull Gray top surfaces with Insignia White side and undersurfaces. VB-131's original PV-1 equipment was similarly camouflaged when first formed at DeLand, Florida.

Lockheed

the squadron mascot, "Skipper" (the dog), was dognapped and two planes and crews were left behind to recover him. This was accomplished with minimum loss of skin and blood, the rescue crews arriving on Attu two days later. "Skipper", and his shipmates arrived safe on Attu. VPB-131 officially relieved VPB-135 of all duties in the Aleutians on 23 October. During the remainder of the month the squadron was engaged in sector searches and Task Force Coverages.

The 11th Army Air Force, 28th Bombardment Group was scheduled to send their B-24 *Liberators* to bomb Karabu Zaki on 4 November. The *Empire Express* was called upon to act as a diversonary force and draw enemy fighters away from the Paramushiro area. A four-plane strike force was assembled, led by LCdr. Hastreiter. This was VPB-131's first mission over the Kuriles. The target chosen was Torishima Retto, a group of small, rock island about ten miles east of the Kurile Straits. No sooner had the PV-1's arrived when they were jumped by eleven enemy fighters. Lt. Ellingboe's plane was hit in the first attack and dove into the sea, exploded and burned.

In the melee which followed one of the turret gunners, G. B. Hall AOM3C was credited with a "probable" when he knocked the cowling off a *Hamp*. It was last seen smoking and losing altitude. The decoy tactic worked. The Army was able to bomb Karabu Zaki with good results and with a minimum amount of opposition. Having battled their way through a swarm of enemy fighters, VB-131 turned for home. They searched the area where Lt. Ellingboe went in the water but all they found was a burning gasoline slick.

Another diversionary mission was undertaken on 6 November. Four PV-1s attacked the east side of Shimushu, while another four from VPB-136 made a high speed photographic run down the west side of Shimushu and Paramushiro. This photo run was the main thrust of the mission and was badly needed for the planned invasion of the Kuriles. The mission was a complete success. All eight PVs returned safely and hundreds of vital photos were obtained. In the meantime VPB-131 was equally successful in keeping many fighters occupied on their side of the islands and eluding them all.

In November 1944 VP-43 (a PBY squadron) was transferred from Attu and VPB-131 and VPB-136 assumed all the patrol duties in the area.

Late in 1944 the Japanese launched a bizzare weapon for striking at the North American Continent. Bomb-carrying balloons were released at a high altitude where the prevailing wind would carry them across the Pacific. These balloon-bombs would drift over Alaska, Canada and some could conceivably finally fall on the United States. A VPB-131 plane on patrol was the first to make contact with one of these balloons. When first sighted the crew presumed it to be a weather balloon and reported sighting same to their base. They were immediately ordered to destroy it. Imagine the crew's surprise when their gunfire exploded it. Thereafter they approached all such floating arsenals with caution.

Another joint Army-Navy strike was scheduled for 26 December 1944. Army B-24s and B-25s were to strike Paramushiro from high altitude, while 131's PV-1s were to draw enemy fighters off to the south. The PVs went in and stirred up a hornet's nest. However, a foul-up occurred in timing and the Army planes were badly mauled over Paramushiro Straits. The PVs were twelve minutes early or the Army planes twelve minutes late. The SNAFU was never fully explained nor understood. Hard feelings resulted between the two services and each went its separate way after the incident.

The New Year was ushered in by intense training in rocket firing — a prelude to a more intense use of these weapons in the Kuriles. From 5 January to 23 January 1945, the western tip of Agattu was battered with High Velocity Aircraft (HVAR) Rockets. These rockets carried five-inch high explosive heads on 3.25-inch rocket motors, with instantaneous nose fuses and .02 second delay base fuses. It was the opinion of the VPB-131 people that the aircraft rocket was not only a lot slower than specified in the frigid temperatures of the area but also that its speed was inconsistent, making firing from the recommended distance of 1000 yards almost impossible. As a result the range was decreased and the aiming accomplished by trial bursts of tracer fire from the bow guns.

Capt. R. Feiten

LCdr. Rolland Hastreiter, left, skipper of VPB-131 and Lt. George Earle, Exec. Officer. Hastreiter was the type of leader who earned the respect of all his men. A great pilot, aggressive, always out front leading, never in the back directing.

John Dawson

Lt. Curt Merkel, ACI Officer of VPB-131 reviews strike photographs with a crew during a briefing.

On 24 January 1945, VPB-131 made its first rocket attack on the Kuriles. Led by LCdr. Hastraiter, they flew in two, two-plane sections, making landfall about thirty miles north of Cape Lopatka for the attack. Hastreiter made the first rocket run over the target, Kokutan Zaki, at an altitude of 1000 feet. He fired his eight rockets in pairs at a dive angle of ten degrees. The first pair was used for sighting and range. All eight rockets hit the target at the base of the lighthouse. Debris was observed flying in all directions. After completing his rocket run he made a turn to port and returned for a strafing run. The turret gunner observed glass being blown out of the tower.

Lt. Feiten's first run over the target was a dry run. He was unable to fire his rockets because of the close interval between the planes so he made a tight turn to starboard and returned at 500 feet altitude and set up the normal dive angle of ten degrees. He fired his rockets in pairs, one failing to fire. Three pairs hit the target, all six going into the base of the radio towers. Smoke, dust and debris were observed after the attack.

Lt. Earle made two runs over the target. The first at an altitude of 100 feet. The first pair of rockets was fired at 900 yards. Four hit at the base of the radio station. He made a tight turn to port and returned at 500 feet altitude, firing his fourth pair of rockets but these went high and over the target.

Lt. Patton made his run at 250 feet firing his rockets in a ripple salvo. Four rockets fell short but four went into the base of the lighthouse. In all, eighteen hits were registered but the lighthouse and radio towers remained standing. Two fighters were sighted but did not close. Ground anti-aircraft fire was moderate, one PV receiving minor damage.

After the attack, the squadron's first with rockets, LCdr. Hastreiter expressed the opinion that the rockets should be fired closer to the target than the recommended 1000 yards. Also the HVAR was not as destructive against such targets as reported, although it was admitted that the high speeds attained during the attacks made assessment of the damage difficult. Later it was determined that the targets remained standing despite the large number of hits scored.

Four PV-1s set out to attack the early warning radar stations at Kurabu Zaki on 2 February. All were on instruments most of the way to the target but Lt. Ellzey became separated from the others. As a result, three made landfall fifty miles north of Cape Lopatka and Lt. Ellzey on Shimushu. All four proceeded to their respective targets.

LCdr. Hastreiter made his approach to Kurabu Zaki somewhat to the south of the radar installations to take advantage of the blocking characteristics of the high terrain. As soon as he determined the target location he turned sharply north and fired all eight rockets. In the pull out and a tight turn to starboard the crew spotted a formation of about fifty Japanese troops marching in formation. They immediately made a strafing run; every gun opening up in a turkey shoot. Many troops were hit and others scattered in frantic disarray.

Lt. Newby followed the C.O. over the target about ten seconds later. His first pair of rockets was fired from an altitude of 1000 feet and were seen to go through the radar screen. The second and third pairs went over the target, but the fourth scored a direct hit at the base of the radar screen. Newby felt the PV shudder as he pulled-up over the hill behind the target. Upon returning to Attu it was found that eighteen inches of the starboard wing tip were torn and a large chunk of Japanese soil was lodged in the tear. The

"The Flying Bulldozer".
The PV-1 of Lt. Don Newby of VPB-131.
They returned to Attu with a chunk
of Japanese soil lodged in the tear.

A. A. Hoffman, AOCS

HEDRON crew presented the chunk of Japanese soil to Lt. Newby and named his PV-1 "The Flying Bulldozer".

Lt. Dawson's run over the target was made at only 300 feet altitude. In attempting to fire a machine gun sighting burst, he inadvertantly fired his rockets instead. As a result they fell short of the target. He pulled up and made a tight turn to starboard, then returned for a strafing run. He credited himself with one snow sled (class unknown) and one horse. Later he mused, "Equipped with chin guns and DOG-1 radar, I'll match the PV-1 against any horse any day".

Because Lt. Ellzey had become separated from the others, he made his run over the fishery at Minami Zaki. They scored two hits with the first salvo of rockets but the last three pairs went over the target.

The target for 8 February was Minami Zaki with Kokutan Zaki as the secondary target; LCdr. Hastreiter again leading the strike. The four PVs maintained visual contact throughout the flight to the target. Minami Zaki was engulfed in frontal activity and snow squalls and the course was changed to the secondary — Kokutan Zaki. The PVs fired their rockets at the lighthouse scoring several direct hits and then made a strafing run.

Enemy shipping became the target for the mission of 20 February. The four-plane foray, lead by Lt. Newby, made landfall about fifty miles south of the mountains of Kamchatka. No shipping was located so the group proceeded to Minami Zaki. Lt. Newby made his first run over the target at 600 feet altitude. The first and second pairs of rockets went over the target and the third and fourth would not fire. Lt. Ellzey made his run at 1000 feet altitude and was a bit more successful. Firing his rockets in pairs, hits were scored on the cannery building. Lt. Powers followed Ellzey. He fired his rockets in pairs, scoring hits on the buildings and the radar screen adjacent to the cannery.

Five minutes after leaving the target, Lt. Powers reported on VHF that he had been hit by flak and had a run-away prop. on the port engine. He feathered the prop. and began making preparations for a landing in Russia. His shipmates immediately joined up and found the crew jettisoning all loose and unnecessary gear. The three PVs escorted Lt. Powers' crippled ship along the coast of Kamchatka toward Petrovalosk. However, Lt. Dawson, in probing the weather toward Petropavlosk, was unable to break out of the fog at 200 feet. The group turned back down the peninsula toward Cape Lopatka hoping for a break in the heavy fog. They knew there was a lighthouse there, which they hoped was manned and could provide friendly aid to Powers and his crew. The escorting planes were forced to return to Attu at this time however, as their fuel was running low. When last seen, Lt. Powers was fifty miles south of Petropavlosk and heading south.

An airman's prayer was answered, the lighthouse at Cape Lopatka was sighted but the rough terrain precluded any attempt at landing. Lt. Powers put the plane on auto-pilot and ordered the crew to bail out. All landed safely, cushioned by a deep snow. Each member was immediately rounded-up by armed Mongolians. To the Americans they looked a lot like Japanese. No English was spoken and their heavy winter uniforms revealed no hint of nationality. They were marched to an underground building where, for the first time, they saw a red star on an officer's uniform . . . they breathed more easily. The crew was transported across Siberia by train and truck and eventually returned to the United States via Europe.

A. A. Hoffman, AOCS

VPB-131, PV-1 BuNo. 49654, flown by Lt. John Powers with port engine damaged over Minami Zaki, heads for Petropavlosk, Russia. The crew had to eventually bail out over Cape Lopatka, on the southern most tip of the Kamchatka Peninsula.

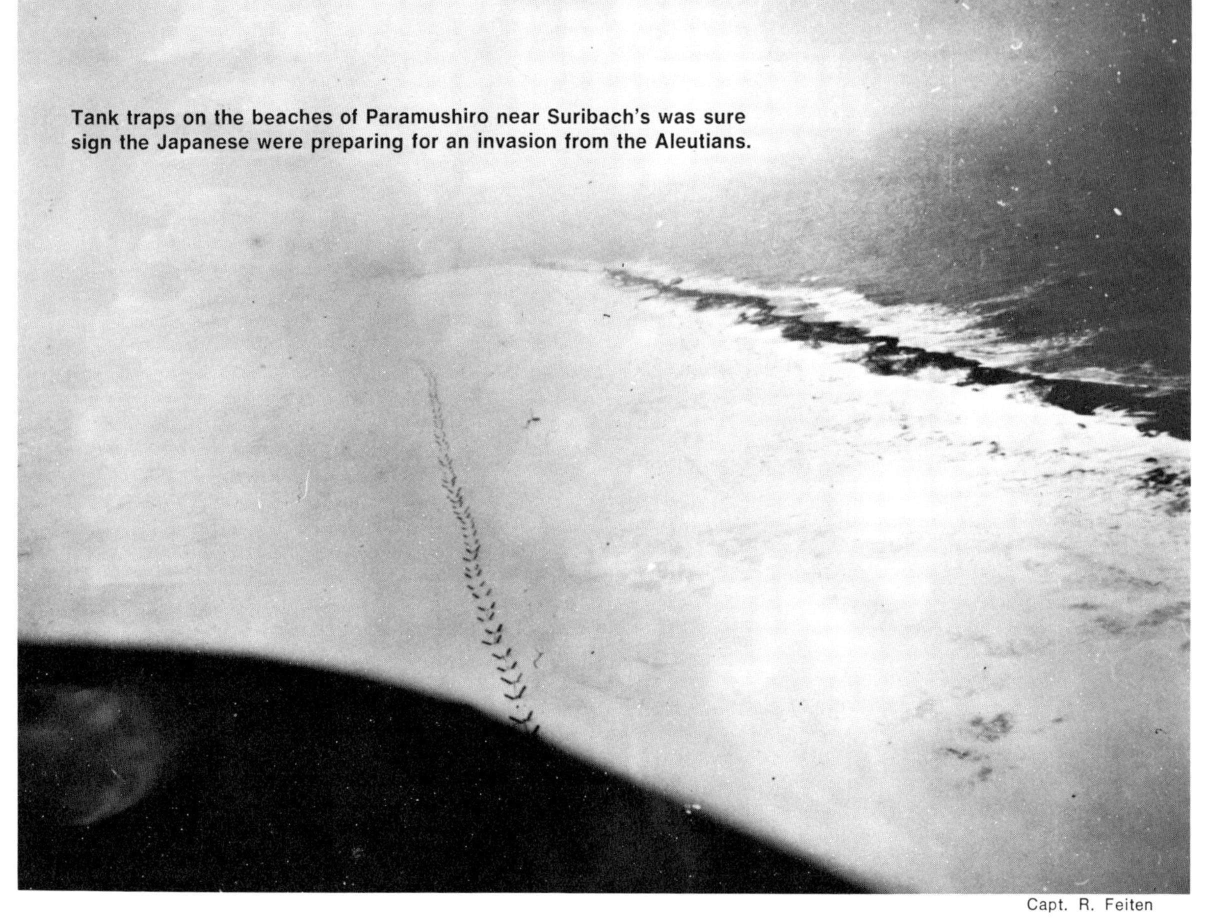

Tank traps on the beaches of Paramushiro near Suribach's was sure sign the Japanese were preparing for an invasion from the Aleutians.

Capt. R. Feiten

Later the photos taken of Lt. Power's plane during his rocket run, showed evidence that the damage they received was probably caused by flying debris from his own rocket hits rather than enemy flak. In essence, Powers, had shot himself down.

The fisheries at Maugawa, Paramushiro were the objective on 23 February. The four PV-1s in the strike force arrived over enemy territory about 100 miles south of the mountains of Kamchatka. No enemy shipping was encountered and with Lt. Berkely in the lead, the group proceeded to the target. Because of the steep cliffs immediately behind the fisheries at Masugawa and the sharp, upward slope beyond, they were unable to make complete and effective runs.

An attack was set up in column form, each plane following the other at 1000 yard intervals. Lt. Berkeley fired his first and second pair of rockets while in a steep bank to avoid hitting the cliffs and consequently made no hits at all. He executed a tight turn to port and returned for his second run, firing his third and fourth pair at a range of 600 yards. Four hits were scored on the buildings. Lt. Brown fired his eight rodkets in salvo but they went wild because he was forced into a sharp bank just as he came on target. Lt. Warnock's rockets went over and exploded far beyond the target. Lt. Barber experienced the same difficulties as the others, with rockets falling short of the target. The whole mission was an exercise in futility.

A single-engine fighter was sighted by Lt. Brown about eight miles east of Masugawa on their return to Attu. It disappeared into the overcast. Four other single-engine aircraft were sighted by Lt. Barber about eight miles southeast of Arahata Zagi but they also slipped into the clouds.

During the month of March the squadron participated in four rocket strikes and two Task Force coverages. This entailed a constant anti-submarine and anti-aircraft screen over the task force on its way in and out of Kuriles area. The strike pattern was essentially unchanged from that used the previous two months, an approach on the deck in single file, followed by a pull-up and final run which was varied to suit the location of the target. The number of missions flown during March was curtailed by the weather.

On 17 March an anti-shipping sweep and attack on Hayake Gawa cannery was scheduled. Four PV-1s tookoff from Attu between 0902 and 0905. Weather conditions were overcast with snow squalls. All planes broke out on top of the overcast and were then able to maintain visual contact. Unfortunately, shortly thereafter three of the planes had to turn back due to mechanical and radio diffculties, Lt. Earle proceeding to the target alone. He made visual landfall 130 miles south of Cape Shipunski. Dropping to the deck he approached Paramushiro but Hawake Gawa was completely socked-in by snow squalls. Due to poor visibility and constant snow flurries, he passed over Torishima Retto without knowing it. Turning to port when he saw the surf off Tomari Zaki, he circled and returned to make his rocket run on the fishery on Torishima Retto.

He opened up with the bow guns, noting that his first burst went just over the buildings. He quickly adjusted aim and fired his rockets in a ripple salvo. The rockets hit "dead on" among the buildings and fishing boats near the shoreline, scoring eight hits with eight rockets. Particularly noted was part of a fishing boat flying high into the air.

Hayake Gawa cannery was the primary target again on 19 March. It was one of those rare CAVU days and Lts. Barkeley, Newby, Warnock and Brown were able to maintain visual contact through the entire mission. No enemy shipping was observed during their initial thrust into the area and with Lt. Berkeley in the lead, the four *Venturas* proceeded to hit the deck for the run on the target.

Lt. Berkeley made the first run, firing his rockets in ripple salvo. He made a wide turn to port, giving the other planes

Burial services for VPB-131 crew, killed when they crashed at Casco Cove, victims of a typical 100 mph Aleutian Willi-waw.

time to make their runs. He then returned for a strafing run. Since the target was obscured by the terrain on his first run, Lt. Newby was unable to get into position for a rocket attack, but the turret gunner fired 150 rounds into the buildings as they passed over the cannery. He then made a steep climbing turn to port, which carried him about one mile further west of the target. Returning he fired all four pairs of rockets, hitting the buildings and several fishing boats.

The rough terrain also made Lt. Warnock's run ineffective. His first pair of rockets scored hits, but the next three were over the targets. Lt. Brown also encountered difficulty in making his run. All his rockets went into the water. He returned for a strafing run, scoring hits on the buildings. Having expended their heavy firepower, the group headed for home.

Three days later the canneries at Asahigawa and No Naka Gawa were again the targets of the *Empire Express*. The four PVs tookoff under typical Aleutian weather conditions; fog with overcast but were unable to rendezvous due to squall conditions and loss of VHF communications. Lts. Earle and Patton had to return to Attu after unsuccessful attempts to communicate with others.

Lts. Feiten and Renshaw continued, maintaining visual contact throughout most of the flight to the targets. Immediately after making landfall they lost sight of the coast line due to fog and haze, but pinpoint navigation brought them directly over the target area. They slowly broke through the fog layer as they lost altitude for the attack. Lt. Feiten made three rocket and two strafing runs. On the first run two pairs of rockets were fired at No Naka Gawa, then banking steeply, he returned for a strafing run. He then turned wide to port, returning for a third run, firing his third pair of rockets. His fourth run was made over Asahigawa. He was unable to observe clearly the results of the attack. Other than that he saw them hit in the general area of the buildings and boats at No Naka Gawa.

Lt. Renshaw made his first run over No Naka Gawa, firing his first two pairs of rockets. Turning to starboard he returned and fired his third and fourth pairs; then experiencing the same difficulties as Lt. Feiten, ceiling and visibility hindering observation of his hits. Immedately after, the two PVs pulled up into the overcast, went on instruments and headed for home.

The officer's mess at NAS Attu, always a pleasant respite from an otherwise bitter cold, miserable and frustrating existence.

John Dawson

The target for 26 March was the cannery at Tomari Zaki. Lt. Berkeley damaged his tail wheel while taxiing and was unable to takeoff. Lt. Barber returned shortly after becoming airborne due to a malfunction in the plane's de-icer boots. When finally they formed-up LCdr. Hastreiter found himself leading only a two plane force to the target area.

Reaching the enemy's cannery without further difficulty, the pair made their attack without hesitation. LCdr. Hastreiter's first rockets appeared to fall short of the buildings but his second and third pairs scored hits and debris was seen flying in all directions. Wingman Lt. Brown had his first pair fall short and tried a second time. Hits were scored this time and flying debris was clearly observed. As they were leaving the area a single *Oscar* was sighted by Brown. Both planes dropped to the deck and increased their speed. After flying parallel for nine minutes, the fighter turned to port and disappeared.

With the arrival in March of VPB-139 and their PV-2 *Harpoons*, it was decided that VPB-131 would be assigned sector searches only. The greater range of the PV-2s made them more suitable for future rocket attacks on the far flung Kuriles. After mid-April, the gas load put aboard the PV-1s for sector searches was reduced from 1620 to 1420 gallons. This was done after the advantages of the extra fuel were weighed against the disadvantages of the extra weight. This also reduced takeoff weight to a more acceptable 32,000 pounds.

One of the more tragic losses of the *Empire Express* occurred on 7 April when Lt. (jg) Patton and his entire crew were killed. They crashed into Casco Cove while attempting to land into the teeth of an Aleutian Willi-Waw. With a 60 knot wind tumbling down the lee side of the mountains, the turbulence was bone-jarring. "Pat" made an approach, found it too rough and started to go around for another try. He almost had it made when a gust flipped him over and the plane went into the cove on its back. The water was so rough at the time that it was almost solid foam. A memorial service was held two days later with the entire squadron in attendance. All members of the crew were buried in the Little Falls Cemetery on Attu.

During May and June, the squadron flew a total of two hundred searches without incident. The large increase in the number of flights was due to more stable weather and the accompanying decrease in frontal activity.

One of the last strikes carried out by VPB-131 against Paramushiro occurred on 5 June. LCdr. Hastreiter and Lt. Berkeley went on a shipping reconnaissance of the Paramushiro Straits. After a negative search of the area the two PV-1s went in on the deck for a closer look. Eight miles east of Suribachi they turned north and paralleled the coast. Skimming the water past Tomari Zaki, they sighted a tug, towing barges toward the Paramushiro Straits. Climbing to 2,000 feet, they made several strafing runs on the tug. Considerable damage to the superstructure was observed but they were unable to determine if the tug was sunk. The squadron made a number of subsequent sorties but missions over the Kuriles became less frequent.

During July the number of searches flown was less than fifty. Increased fog and low cloud layers made the danger of the field closing while aircraft were on patrol more and more prevelent. On more occasions than the crews wanted to remember they owed their lives to the assistance of GCA. Flying in the Aleutians wasn't very forgiving.

It was 2 August 1945, that the big four engined PB4Y-2 *Privateers* of VPB-120 arrived on Attu. VPB-131 was relieved of duty upon their arrival and began departing for the United States.

It was on 6 August 1945, the day that the first atom bomb was released over Hiroshima, that eleven war-weary PV-1s and their tired crews landed at NAS, Whidbey Island, Washington. The twelfth *Ventura* was delayed by mechanical trouble and arrived three days later, only to hear the news that the second A-bomb fell on Nagasaki. The war was near an end. VPB-131 had experienced the longest tour of duty in the Aleutians of any *Empire Express* squadron. They had lost three planes and crews including Lt. Powers and crew, who were forced down in Russia. Certainly they had done their share in bringing the enemy to his knees.

Lt. "Flying Bulldozer" Newby and crew. Back row, l. to r.: Lt.(j.g.) Brock, Lt. Newby, Ens. Overton. Front row: Panetti, Tanner, Grimes. Large thermos jug contained coffee — what else? (always with sugar), cardboard box filled with flight rations — Bologna sandwiches and either apples or oranges.

All Photos John Dawson

VPB-131 crew, Lt. Robert Feiten PPC. Back row, l. to r.: Feiten, Sturgess, Lapham. Front row: Dunn, Keys, Nichols. Below, Lt. John Dawes and crew, back row, l. to r.: Mantz, Dawson, Suddarth. Front row: Santowski, Jones, Foster.

THE Admiral's Cow

USN via L. A. Patteson

The cow's arrival on Attu. After unloading from the USS Avocet, she was the only lady on the island.

No story of the PV squadrons' exploits in the Aleutian area would be complete without the story of the Admiral's Cow. After the capture of Attu, the Admiral in command of Fleet Air Wing Four decided to move his headquarters from Adak to Attu. He had the SeaBees build a suitable headquarters for him and his staff adjacent to the air strip at Murder Point. The ensuing structure was referred to as "The Palace" by the men of FAW-4. The Admiral soon decided that one of the comforts of home, missing in their drab existance, was fresh milk. He, and several thousand other World War II American servicemen, did not like the dried milk of the period. It would not stay in solution. Stir as you would before you drank it, it would definitely settle to the bottom of the glass where it looked and tasted like chalk. It was reasoned that real milk could only be obtained from a real, genuine, living cow. Cows were as scarce as humming birds' nests in the Aleutians. Native Aleuts had never even seen one! Two factors prevented their propagation on the tree-less islands; the severe cold winter and lack of suitable forage. However, through the "can-do" ingenuity of the SeaBees, no obstacle was too difficult to overcome.

The seaplane tender USS *Avocet* (AVP-4) was detached to Attu in support of the PBY squadrons in the area. The Admiral solicited contributions from the members of his staff. One of his aides went aboard the *Avocet* when she sailed for Seattle. The aid's qualification for this assignment was that he was originally a farm boy. The Admiral ordered him to purchase a *good* cow. The ship had been in the cursed cold area for some time and merriment reigned aboard. This was the answer to the crew's prayers — the prospect of liberty in the United States and GIRLS! GIRLS! GIRLS! After what seemed like an eternity, the *Avocet* finally returned to Murder Point. It is doubtful if a stranger looking Man-of-War ever steamed up Puget Sound and one wonders at the comments of those who saw her . . . with a *cow* tied on the fan tail and the hangar full of hay.

While the *Avocet* was on her supply mission to Seattle, the admiral had the SeaBees build a barn, complete with barnyard but at a suitable distance from the "Palace", so that the odors would not permeate the immediate area. Unfortunately, the only suitable location was in close proximity to the quonset huts of the PV squadrons.

The farmer turned sailor was well qualified for his assignment. The cow he purchased gave more milk than the Admiral and his staff could possibly use. He then prevailed upon the officers mess to use the surplus. Needless to say, they complied. However, they had seen the sanitary conditions at the barn and some of them had serious misgivings about the processor's ability to meet Grade A standards. One VB-135 pilot came up with the perfect solution. They would make "Alexanders" from the milk, reasoning that the alcohol would kill any germs present in the milk.

Eventually the Admiral was transferred to a carrier and his replacement was not interested in the cow. Early one morning the squadron personnel was awakened by a bugle call down at the barn. They poked their heads out of their quarters in time to see the admiral's staff, resplendent in full-dress uniforms, drawn up in formation at the barnyard. One of the aides was reading the orders transferring the cow from Fleet Air Wing Four to an Army engineers' outfit. The Army outfit slaughtered the cow and the saga of the Admiral's Cow came to an end.

The cow and her happy home, complete with electric lines for lighting and heat. It was reported that an Army Colonel offered to trade quarters with the bovine.

USN via L. A. Patteson

VB-139, PV-1 taxiing for takeoff from frigid snow covered runway at Casco Field, Attu, Feb. 1944. VB-139 pioneered night photorecon flights over the Kuriles.

U.S.N.

Early PV-1 as delivered to VB-136 at Whidbey Island in early 1943, two-tone Blue-Gray over Light-Gray. Cartoon was applied at the factory.

PV-1 on company check flight prior to delivery to the Navy.

PV-1 Ventura, BuNo. 34986 on flight tests at NAS Patuxent River, Md. 3-31-44.

Lockheed PV-1 Ventura

The dependable P & W R-2800-31 engine; power that helped make the PVs the great success they were. It was rated at 2000 HP at 2700 rpm and 52 inches of manifold pressure, but with throttles firewalled over the Kuriles, the ram-effect of the cold air at sea level resulted in 2200 HP at 56 inches.

F. A. Henninger

Lockheed

"Chin-Gun" package increased forward firepower from 2 to 5 .50 cal. guns. When boresighted correctly, a pilot could literally cut a small ship in half.

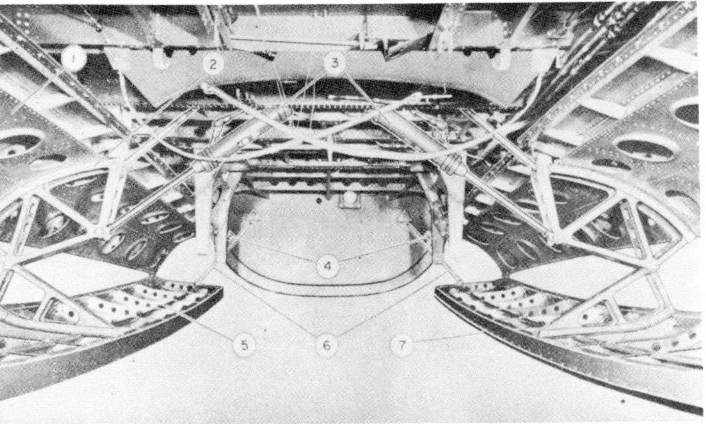

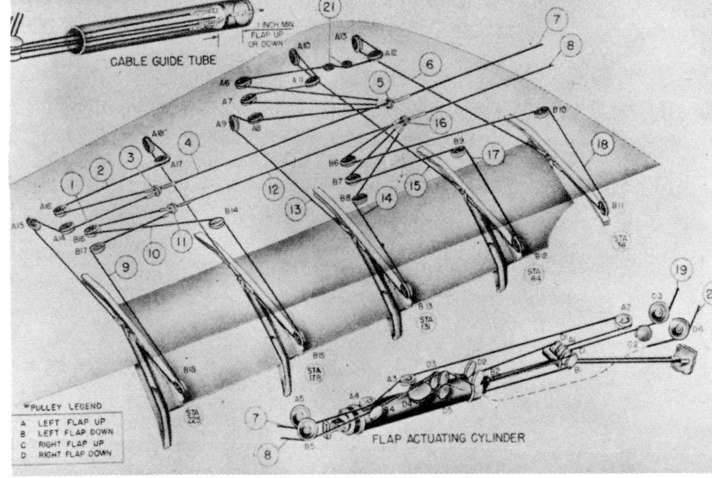

Above, PV bomb bay interior. Right, Fowler flaps details.

L. A. Patteson

Typical art work on VB-135 PV-1s.

L. A. Patteson

Capt. R. Feiten

Woody Keys, AMM/c, Lt. Feiten's radar operator/tunnel gunner, poses alongside his art work on their PV-1. This is the same plane in which Lt. Powers and crew bailed out of over Russia.

FLEET AIR WING 4

The Empire Express — Flying into the Japanese Empire through the worst of weather, day and night, with bat like canniness.

COLORS: Red "Rising Sun" on white background. Bat silhouette — black. Ocean — dark blue with white wavy scroll decor.

This insignia is believed to have been placed centrally on the fin on only a few PV-1s and possibly by VB-135 only.

VB-135: The blue fox, the only native inhabitant of the Aleutians, rides a flying gas tank, (the PV-1). The blindfold indicates the flying conditions encountered and the cane suggests the radar used to find the target.
COLORS: Fox and mountain — pale blue. Clouds, snow capped mountain, blindfold, gun package, fox's bib and tip of tail — white. Large bomb — dark green. Gas tank, wings, belt and neck life preserver — yellow. Outlined and detailed in — black.

VB-136: Insignia during first tour of duty in 1943. "N" on rabbit's shirt stood for LCdr Nathan "Nate" Haines, the squadron skipper. Lt. H. H. Throckmorton, a VB-136 pilot, designed the insignia. He was killed in a crash of a PV-1 on takeoff from Adak.

COLORS: Unknown

VB-136: Insignia during second tour of duty — 1944-45. LCdr. Charles Wayne, Skipper at the time, liked the Husky dog. This insignia had originally been used by VP-41, a PBY squadron who was the predecessor of VB-136.

COLORS: Border — red. Sky — blue. Trees — green with patches of white snow. Ground area — white. Dog — brown and white.

VB-139: Cigar smoking yellow-jacket gave deadly sting to VB-139s operation covering the top of the world.

COLORS: Bee — black and yellow, with yellow blending into inner disc area. Remaining colors unknown.

VPB-131: The winged gauntlet of challenge. A special Disney Studio design for VPB-131.

COLORS: Gauntlet and forewing — yellow ocre. Right or background wing — light olive green. Background disc, sky — dark blue. Ocean — bright green.

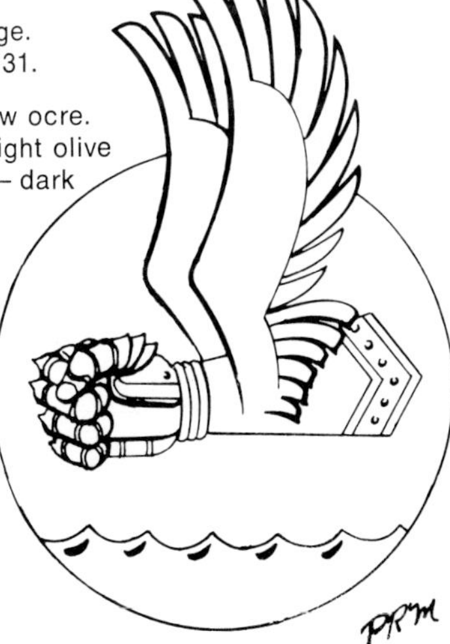

USN via L. A. Patteson

Instrument panel and controls of a PV-1 Ventura while in flight.

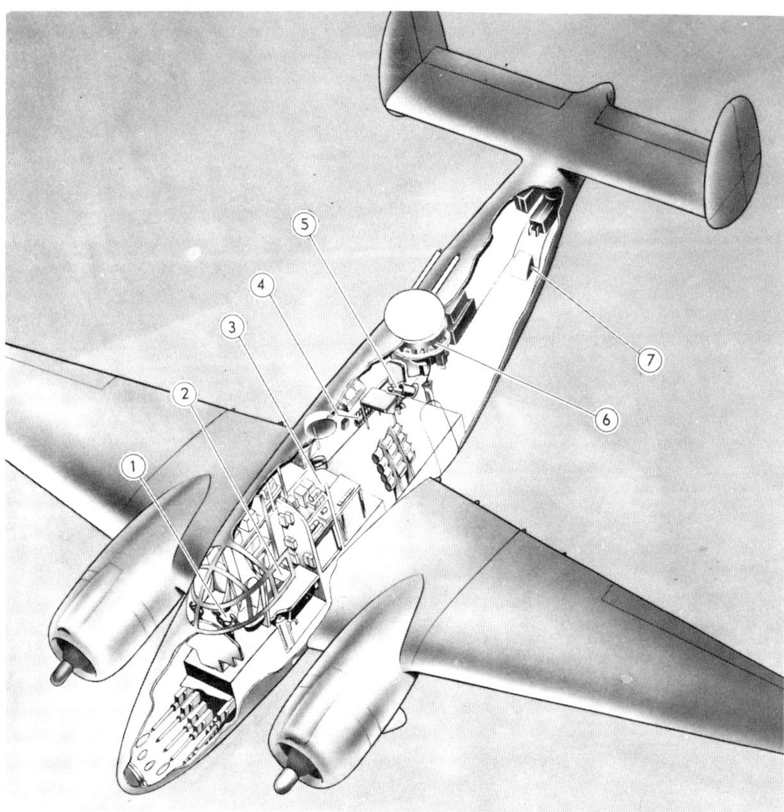

Lockheed

PV-2 cut-away. 1-Pilot, 2-Radioman, 3-Navigator's table with fuel tank underneath, 4-Radarman's table, 5-Flare chute, 6-Martin electric turrett (twin .50 cal. guns), 7-Tunnel.

A. A. Hoffman, AOCS

Interior of PV-1, looking aft from Radioman's station. Radar set on left, navigator is sitting on oil tank. Light coming between men is from the ventral tunnel gun position.

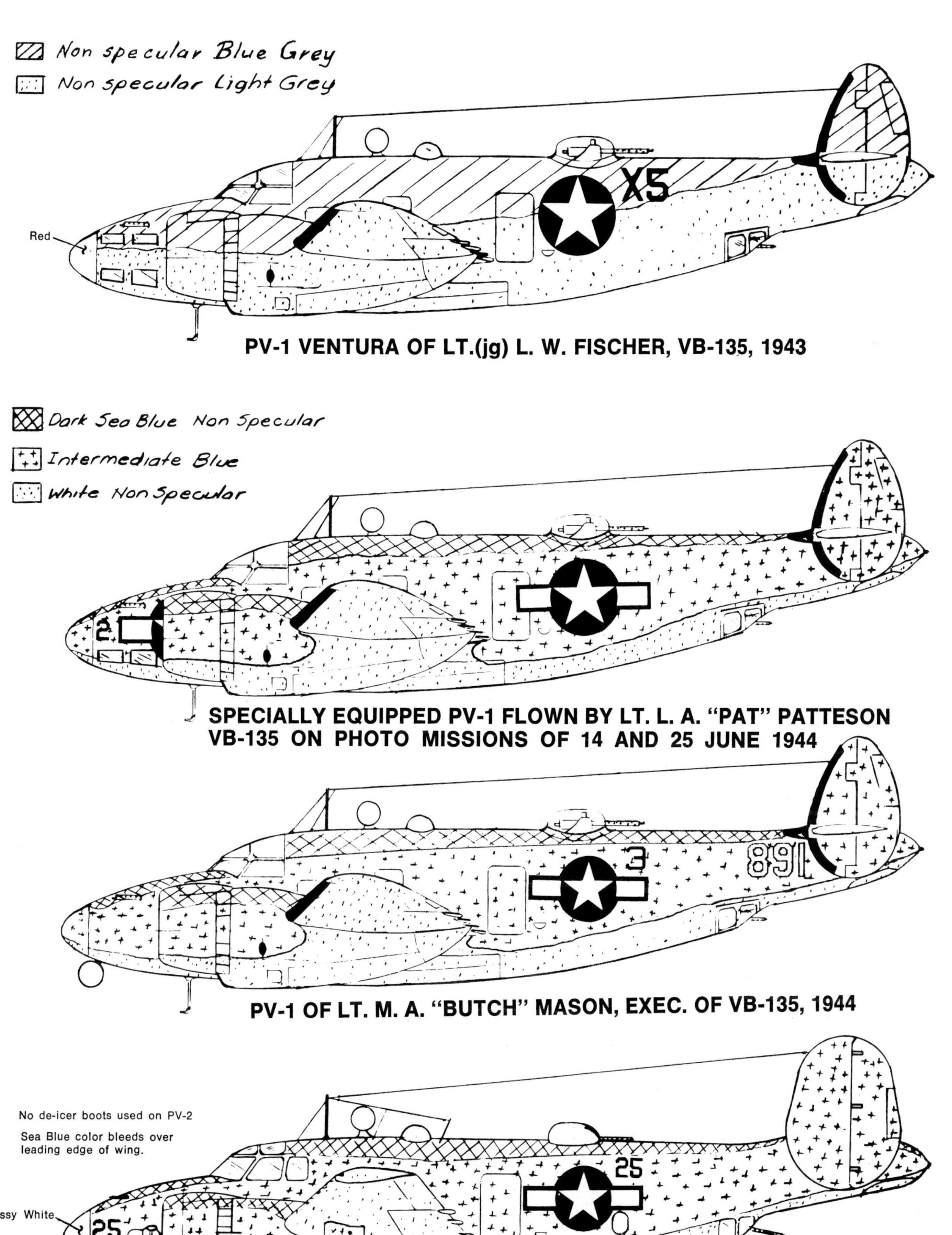

PV-1 VENTURA OF LT.(jg) L. W. FISCHER, VB-135, 1943

SPECIALLY EQUIPPED PV-1 FLOWN BY LT. L. A. "PAT" PATTESON VB-135 ON PHOTO MISSIONS OF 14 AND 25 JUNE 1944

PV-1 OF LT. M. A. "BUTCH" MASON, EXEC. OF VB-135, 1944

PV-2 HARPOON OF LCDR. GLENN DAVID, SKIPPER OF VPB-139, 1945

Drawings by Charles Scrivner

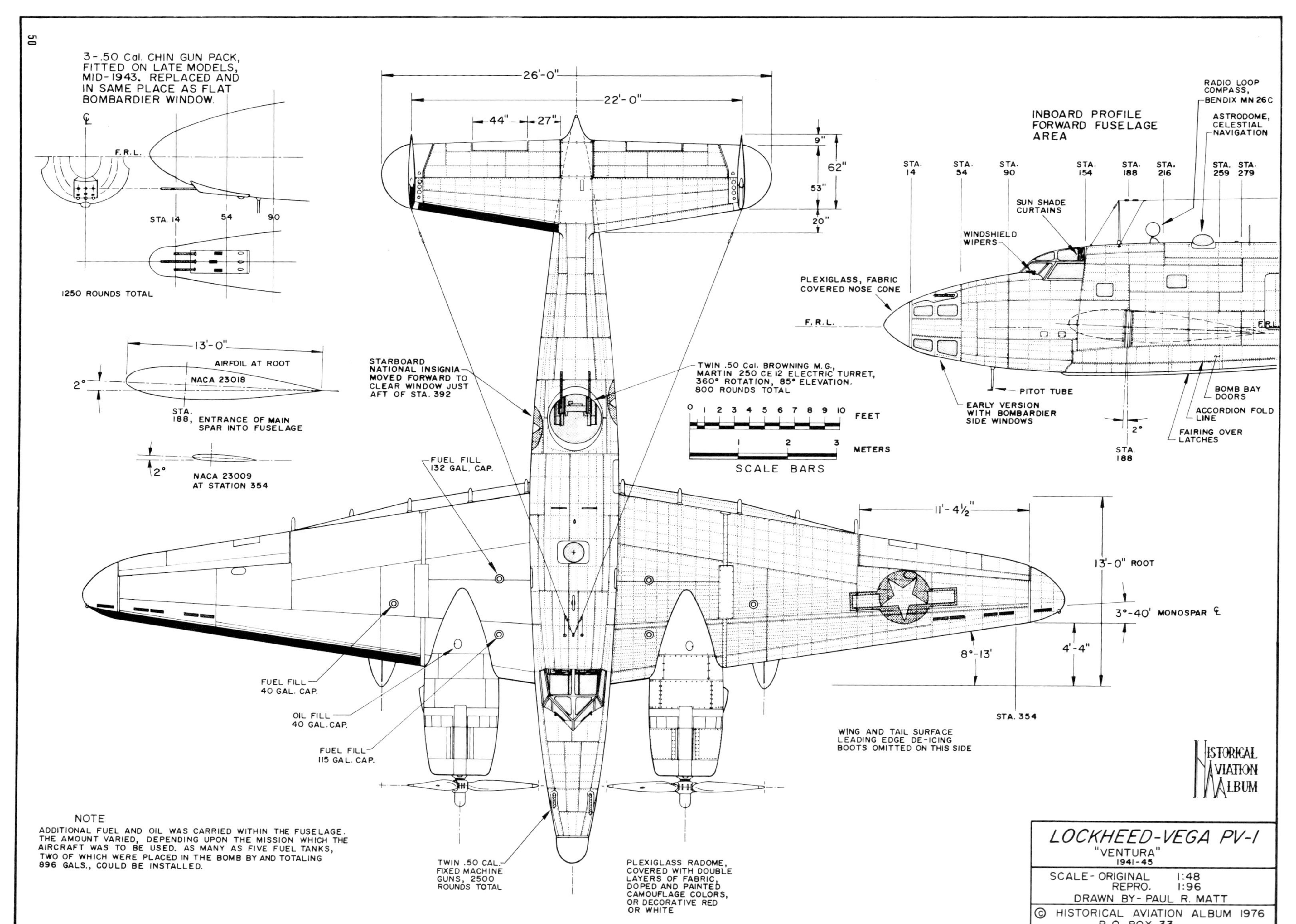
3-.50 Cal. CHIN GUN PACK, FITTED ON LATE MODELS, MID-1943. REPLACED AND IN SAME PLACE AS FLAT BOMBARDIER WINDOW.
F.R.L.
STA. 14
54
90
1250 ROUNDS TOTAL
26'-0"
22'-0"
44"
27"
9"
62"
53"
20"
13'-0"
AIRFOIL AT ROOT
NACA 23018
2°
STA. 188, ENTRANCE OF MAIN SPAR INTO FUSELAGE
NACA 23009 AT STATION 354
STARBOARD NATIONAL INSIGNIA MOVED FORWARD TO CLEAR WINDOW JUST AFT OF STA. 392
TWIN .50 Cal. BROWNING M.G., MARTIN 250 CE 12 ELECTRIC TURRET, 360° ROTATION, 85° ELEVATION. 800 ROUNDS TOTAL
0 1 2 3 4 5 6 7 8 9 10 FEET
1 2 3 METERS
SCALE BARS
FUEL FILL 132 GAL. CAP.
FUEL FILL 40 GAL. CAP.
OIL FILL 40 GAL. CAP.
FUEL FILL 115 GAL. CAP.
TWIN .50 CAL. FIXED MACHINE GUNS, 2500 ROUNDS TOTAL
PLEXIGLASS RADOME, COVERED WITH DOUBLE LAYERS OF FABRIC, DOPED AND PAINTED CAMOUFLAGE COLORS, OR DECORATIVE RED OR WHITE
INBOARD PROFILE FORWARD FUSELAGE AREA
STA. 14
STA. 54
STA. 90
STA. 154
STA. 188
STA. 216
STA. 259
STA. 279
RADIO LOOP COMPASS, BENDIX MN 26C
ASTRODOME, CELESTIAL NAVIGATION
SUN SHADE CURTAINS
WINDSHIELD WIPERS
PLEXIGLASS, FABRIC COVERED NOSE CONE
F.R.L.
PITOT TUBE
EARLY VERSION WITH BOMBARDIER SIDE WINDOWS
BOMB BAY DOORS
ACCORDION FOLD LINE
FAIRING OVER LATCHES
2°
STA. 188
11'-4½"
13'-0" ROOT
3°-40' MONOSPAR ℄
4'-4"
8°-13'
STA. 354
WING AND TAIL SURFACE LEADING EDGE DE-ICING BOOTS OMITTED ON THIS SIDE
HISTORICAL AVIATION ALBUM
NOTE
ADDITIONAL FUEL AND OIL WAS CARRIED WITHIN THE FUSELAGE. THE AMOUNT VARIED, DEPENDING UPON THE MISSION WHICH THE AIRCRAFT WAS TO BE USED. AS MANY AS FIVE FUEL TANKS, TWO OF WHICH WERE PLACED IN THE BOMB BY AND TOTALING 896 GALS., COULD BE INSTALLED.
LOCKHEED-VEGA PV-1
"VENTURA"
1941-45
SCALE- ORIGINAL 1:48
REPRO. 1:96
DRAWN BY- PAUL R. MATT
© HISTORICAL AVIATION ALBUM 1976
P.O. BOX 33

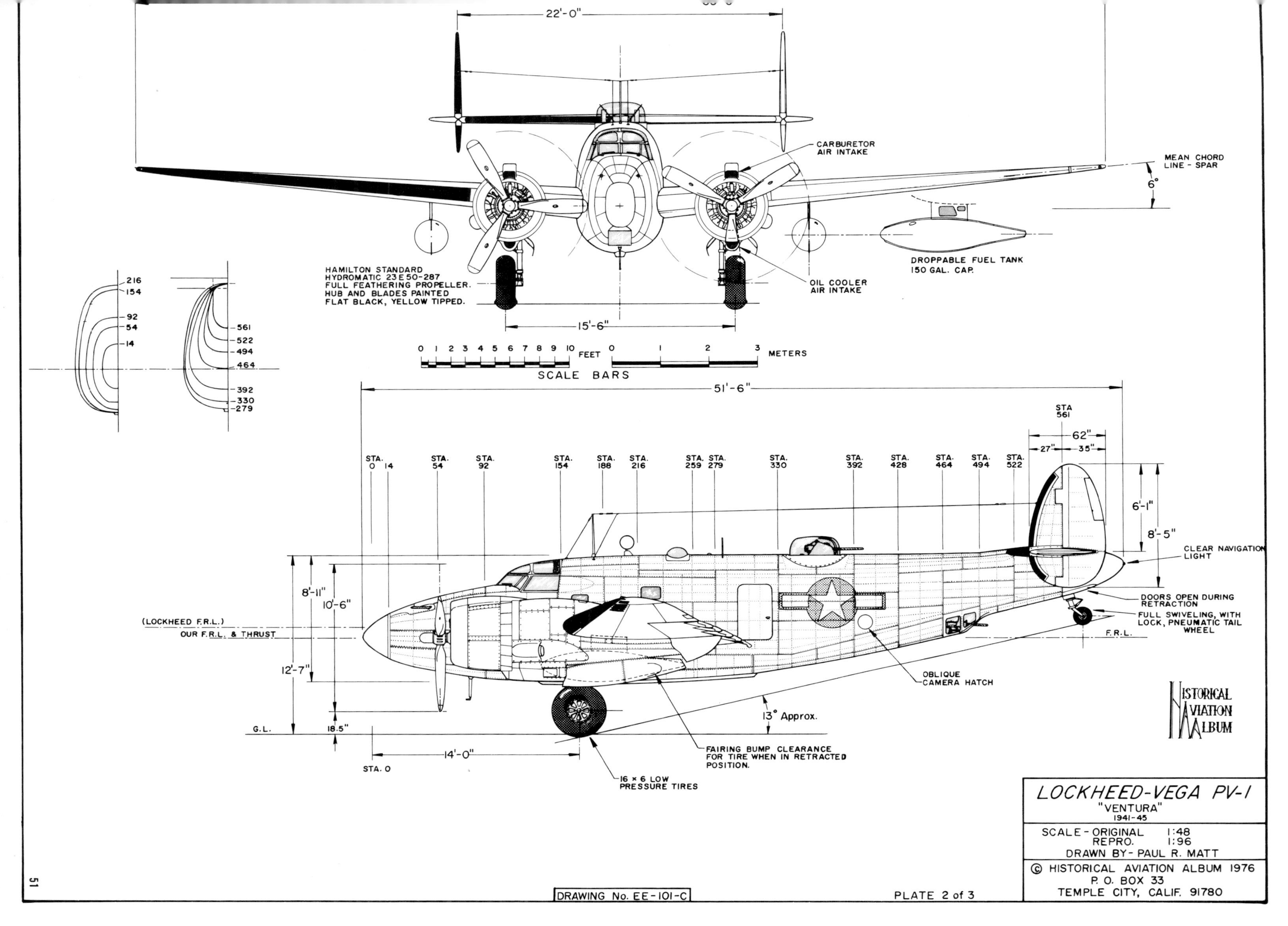
22'-0"
CARBURETOR AIR INTAKE
MEAN CHORD LINE - SPAR
6°
DROPPABLE FUEL TANK 150 GAL. CAP.
HAMILTON STANDARD HYDROMATIC 23E50-287 FULL FEATHERING PROPELLER. HUB AND BLADES PAINTED FLAT BLACK, YELLOW TIPPED.
OIL COOLER AIR INTAKE
15'-6"
216
154
92
54
14
561
522
494
464
392
330
279
0 1 2 3 4 5 6 7 8 9 10 FEET
0 1 2 3 METERS
SCALE BARS
51'-6"
STA 561
62"
27"
35"
STA. 0
14
STA. 54
STA. 92
STA. 154
STA. 188
STA. 216
STA. 259
STA. 279
STA. 330
STA. 392
STA. 428
STA. 464
STA. 494
STA. 522
6'-1"
8'-5"
CLEAR NAVIGATION LIGHT
8'-11"
10'-6"
(LOCKHEED F.R.L.)
OUR F.R.L. & THRUST
DOORS OPEN DURING RETRACTION
FULL SWIVELING, WITH LOCK, PNEUMATIC TAIL WHEEL
F.R.L.
12'-7"
OBLIQUE CAMERA HATCH
13° Approx.
G.L.
18.5"
14'-0"
STA. 0
FAIRING BUMP CLEARANCE FOR TIRE WHEN IN RETRACTED POSITION.
16 × 6 LOW PRESSURE TIRES
HISTORICAL AVIATION ALBUM
LOCKHEED-VEGA PV-1
"VENTURA"
1941-45
SCALE - ORIGINAL 1:48
REPRO. 1:96
DRAWN BY- PAUL R. MATT
© HISTORICAL AVIATION ALBUM 1976
P. O. BOX 33
TEMPLE CITY, CALIF. 91780
DRAWING No. EE-101-C
PLATE 2 of 3

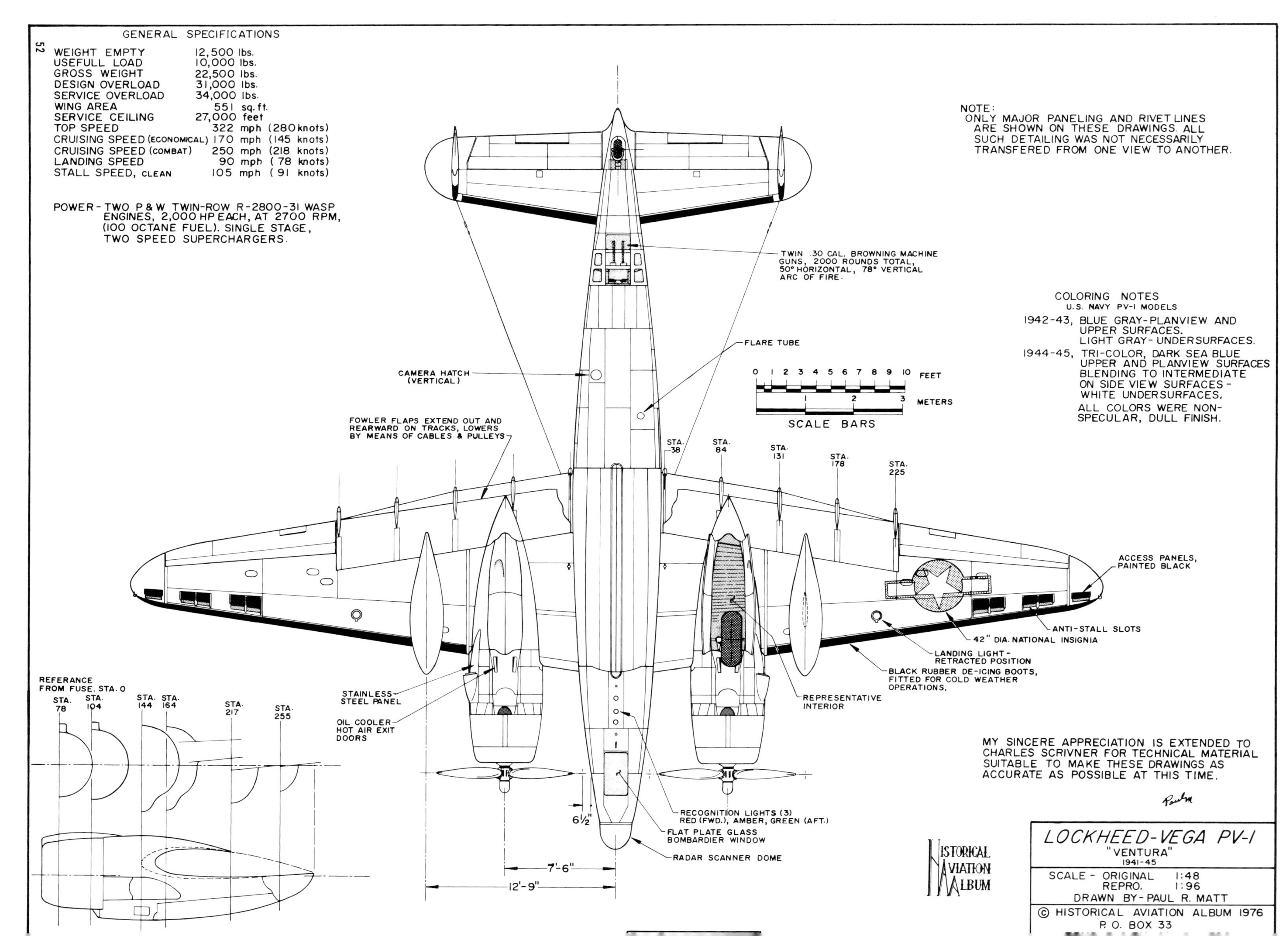
GENERAL SPECIFICATIONS
WEIGHT EMPTY 12,500 lbs.
USEFULL LOAD 10,000 lbs.
GROSS WEIGHT 22,500 lbs.
DESIGN OVERLOAD 31,000 lbs.
SERVICE OVERLOAD 34,000 lbs.
WING AREA 551 sq. ft.
SERVICE CEILING 27,000 feet
TOP SPEED 322 mph (280 knots)
CRUISING SPEED (ECONOMICAL) 170 mph (145 knots)
CRUISING SPEED (COMBAT) 250 mph (218 knots)
LANDING SPEED 90 mph (78 knots)
STALL SPEED, CLEAN 105 mph (91 knots)
POWER - TWO P & W TWIN-ROW R-2800-31 WASP ENGINES, 2,000 HP EACH, AT 2700 RPM, (100 OCTANE FUEL). SINGLE STAGE, TWO SPEED SUPERCHARGERS.
NOTE:
ONLY MAJOR PANELING AND RIVET LINES ARE SHOWN ON THESE DRAWINGS. ALL SUCH DETAILING WAS NOT NECESSARILY TRANSFERED FROM ONE VIEW TO ANOTHER.
TWIN .30 CAL. BROWNING MACHINE GUNS, 2000 ROUNDS TOTAL, 50° HORIZONTAL, 78° VERTICAL ARC OF FIRE.
COLORING NOTES
U.S. NAVY PV-1 MODELS
1942-43, BLUE GRAY-PLANVIEW AND UPPER SURFACES. LIGHT GRAY-UNDERSURFACES.
1944-45, TRI-COLOR, DARK SEA BLUE UPPER AND PLANVIEW SURFACES BLENDING TO INTERMEDIATE ON SIDE VIEW SURFACES - WHITE UNDERSURFACES.
ALL COLORS WERE NON-SPECULAR, DULL FINISH.
FLARE TUBE
CAMERA HATCH (VERTICAL)
0 1 2 3 4 5 6 7 8 9 10 FEET
1 2 3 METERS
SCALE BARS
FOWLER FLAPS EXTEND OUT AND REARWARD ON TRACKS, LOWERS BY MEANS OF CABLES & PULLEYS
STA. 38
STA. 84
STA. 131
STA. 178
STA. 225
ACCESS PANELS, PAINTED BLACK
ANTI-STALL SLOTS
42" DIA. NATIONAL INSIGNIA
LANDING LIGHT - RETRACTED POSITION
BLACK RUBBER DE-ICING BOOTS, FITTED FOR COLD WEATHER OPERATIONS.
REPRESENTATIVE INTERIOR
REFERANCE FROM FUSE. STA. 0
STA. 78
STA. 104
STA. 144
STA. 164
STA. 217
STA. 255
STAINLESS STEEL PANEL
OIL COOLER HOT AIR EXIT DOORS
6½"
7'-6"
12'-9"
RECOGNITION LIGHTS (3) RED (FWD.), AMBER, GREEN (AFT.)
FLAT PLATE GLASS BOMBARDIER WINDOW
RADAR SCANNER DOME
MY SINCERE APPRECIATION IS EXTENDED TO CHARLES SCRIVNER FOR TECHNICAL MATERIAL SUITABLE TO MAKE THESE DRAWINGS AS ACCURATE AS POSSIBLE AT THIS TIME.
HISTORICAL AVIATION ALBUM
LOCKHEED-VEGA PV-1
"VENTURA"
1941-45
SCALE - ORIGINAL 1:48
REPRO. 1:96
DRAWN BY - PAUL R. MATT
© HISTORICAL AVIATION ALBUM 1976
P. O. BOX 33

THE PV-2 HARPOON

Lockheed

Early PV-2 sporting Dark Sea Blue upper surfaces.

Lockheed

PV-2 Harpoon over Alaska.

Lockheed

PV-2 undergoing armament (FT) tests by the Navy.

Cockpit of a PV-2 Harpoon. Visibility was excellent from PVs. Insulation along sides of fuselage was a PV-2 niceity, something the PV-1 did not possess.

U.S. Navy

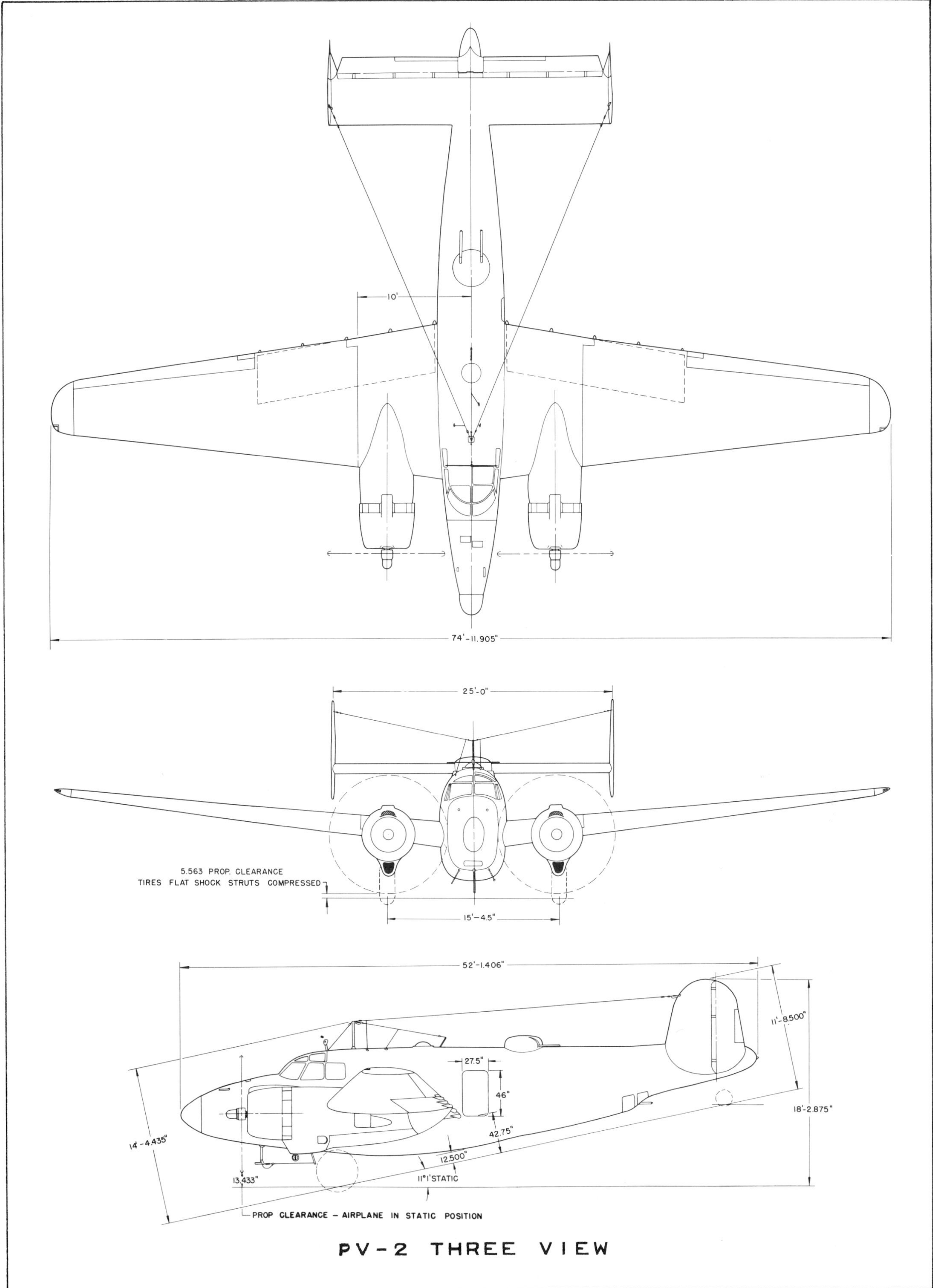

PV-2 THREE VIEW

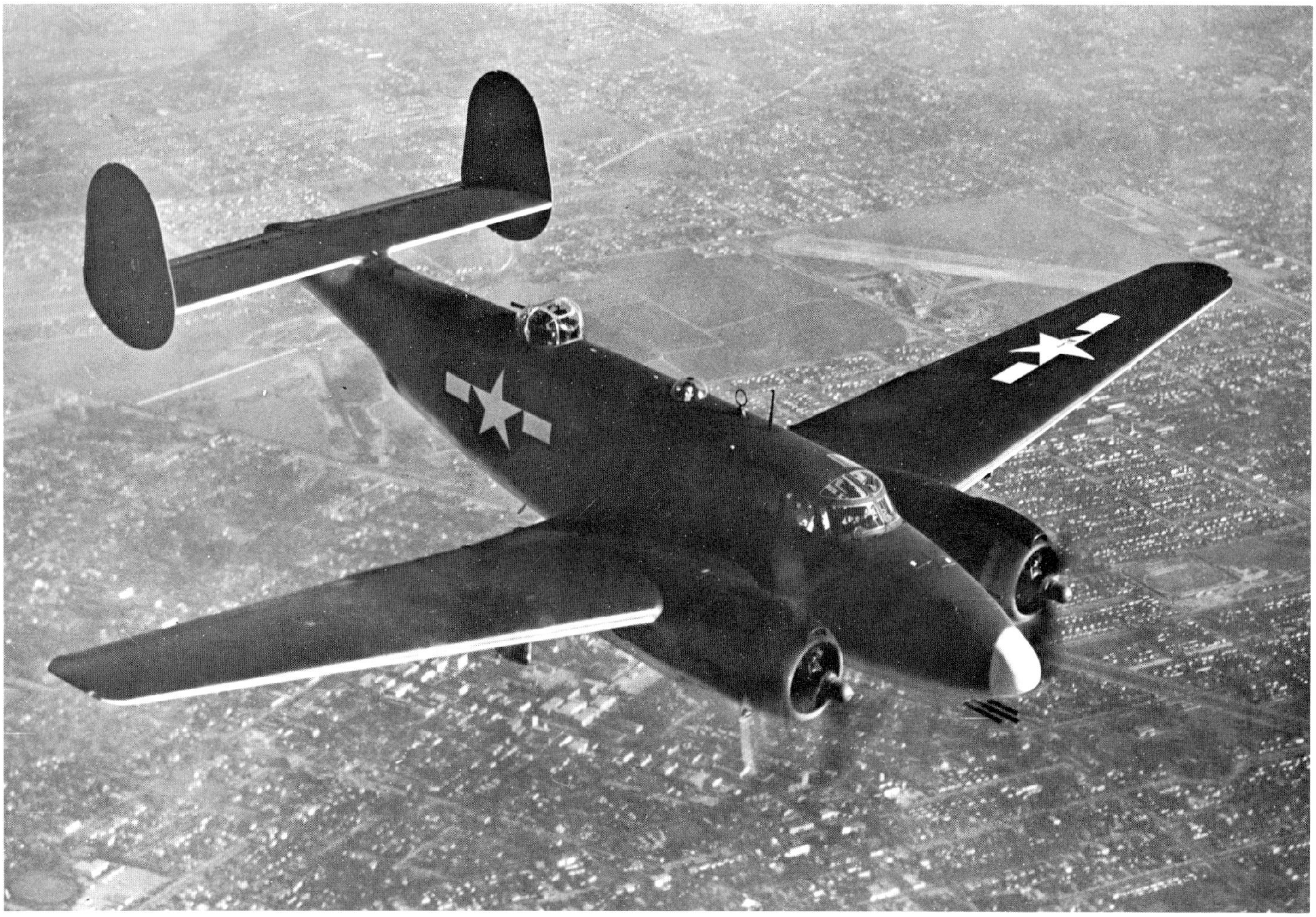

Lockheed

Lockheed PV-2 Harpoon, above, in final service camouflage and, below, in early scheme.

Lockheed